High Performance Computing Symposium

(HPCS 2013)

Journal of Physics: Conference Series Volume 540

Ottawa, Canada
2-6 June 2013

ISBN: 978-1-63439-465-9
ISSN: 1742-6588

Printed from e-media with permission by:

Curran Associates, Inc.
57 Morehouse Lane
Red Hook, NY 12571

Some format issues inherent in the e-media version may also appear in this print version.

Printed by Curran Associates, Inc. (2014)

For permission requests, please contact the Institute of Physics
at the address below.

Institute of Physics
Dirac House, Temple Back
Bristol BS1 6BE UK

Phone: 44 1 17 929 7481
Fax: 44 1 17 920 0979

techtracking@iop.org

Additional copies of this publication are available from:

Curran Associates, Inc.
57 Morehouse Lane
Red Hook, NY 12571 USA
Phone: 845-758-0400
Fax: 845-758-2634
Email: curran@proceedings.com
Web: www.proceedings.com

Table of contents

Volume 540

High Performance Computing Symposium 2013 (HPCS 2013)
2–6 June 2013, Ottawa, Canada

Accepted papers received: 29 August 2014
Published online: 13 October 2014

Preface

011001

High Performance Computing Symposium 2013 (HPCS 2013)

Jonathan Dursi, Nasser Mohieddin Abukhdeir, Mark Daley, Igor Jurisica, Doug Mewhort, Ralf Meyer and Gary Slater

011002

Peer review statement

Papers

012001

A case-comparison study of automatic document classification utilizing both serial and parallel approaches 1

B Wilges, R C Bastos, G P Mateus and M A R Dantas

012002

Towards a Resource Reservation Approach for an Opportunistic Computing Environment 11

E Gomes and M A R Dantas

012003

Modeling Remote I/O versus Staging Tradeoff in Multi-Data Center Computing 20

I H Suslu

012004

Implementation of a solution Cloud Computing with MapReduce model 34

C Baya

012005

Green Cloud Computing: An Experimental Validation 43

R C Monteiro, M A R Dantas and M V Rodriguez y Rodriguez

012006

Efficient parallelization of molecular dynamics simulations with short-ranged forces 54

R Meyer

012007

High-Dimensional Adaptive Particle Swarm Optimization on Heterogeneous Systems 61

M P Wachowiak, B B Sarlo and A E L Foster

012008

Modeling a distributed environment for a petroleum reservoir engineering application with software product line 70

R de F Scheidt, P Vilain and M A R Dantas

Journal of Physics: Conference Series **540** (2014) 011001 doi:10.1088/1742-6596/540/1/011001

The HPCS2013 Proceedings

Jonathan Dursi, Nasser Mohieddin Abukhdeir, Mark Daley, Igor Jurisica, Doug Mewhort, Ralf Meyer, Gary Slater

E-mail: ljdursi@hpcs.ca

The Program committee of HPCS2013 would like to thank those who contributed to HPCS2013, through the technical program, the Birds of Feather sessions, the vendor overviews, the networking sessions, or for attending and grilling the speakers in all of these sessions with great questions and contributing to fantastic discussions.

We'd particularly like to highlight the best paper award presented at the conference, going to "The Making of Big Brain", presented by Marc-Étienne Rousseau for the Big Brain team; the best student paper for "Towards a Resource Reservation Approach for an Opportunistic Computing Environment", presented by Eliza Gomes; and the best visualization, to a movie of an amazing globe-to-individual-building level simulation of the evolution of a toxic plume over a city, presented by Bertrand Denis of the Canadian Meteorological Centre.

It was a great conference, and we look forward to seeing you in Halifax for HPCS2014!

Journal of Physics: Conference Series **540** (2014) 011002 doi:10.1088/1742-6596/540/1/011002

Peer review statement

All papers published in this volume of *Journal of Physics: Conference Series* have been peer reviewed through processes administered by the proceedings Editors. Reviews were conducted by expert referees to the professional and scientific standards expected of a proceedings journal published by IOP Publishing.

Journal of Physics: Conference Series **540** (2014) 012001 doi:10.1088/1742-6596/540/1/012001

A case-comparison study of automatic document classification utilizing both serial and parallel approaches

B Wilges[1], R C Bastos[1], G P Mateus[2] and M A R Dantas[2]
[1] Department of Engineering and Knowledge Management (EGC)
Federal University of Santa Catarina (UFSC)
Florianópolis, SC, 88040-900, Brazil
[2] Department of Informatics and Statistic (INE)
Federal University of Santa Catarina (UFSC)
Florianópolis, SC, 88040-900, Brazil

E-mail: beaw@inf.ufsc.br

Abstract. A well-known problem faced by any organization nowadays is the high volume of data that is available and the required process to transform this volume into differential information. In this study, a case-comparison study of automatic document classification (ADC) approach is presented, utilizing both serial and parallel paradigms. The serial approach was implemented by adopting the RapidMiner software tool, which is recognized as the world-leading open-source system for data mining. On the other hand, considering the MapReduce programming model, the Hadoop software environment has been used. The main goal of this case-comparison study is to exploit differences between these two paradigms, especially when large volumes of data such as Web text documents are utilized to build a category database. In the literature, many studies point out that distributed processing in unstructured documents have been yielding efficient results in utilizing Hadoop. Results from our research indicate a threshold to such efficiency.

1. Introduction

The intensive utilization of the web has created an unbelievable volume of information never seen before. Inside the World Wide Web environment there are three classes of data: unstructured, semi-structured and structured data. The first type of data is represented by the HyperText Markup Language (HTML), semi-structured by the Extensible Markup Language (XML), and, finally, conventional databases can represent structured data. As mentioned in the report from the Royal Pingdom [1], an Internet monitoring company, in December 2011 the number of sites was circa 555 million, with an increase of 300 million added only in 2011. In this year, other interesting examples are the growth of Web servers utilizing Apache technology, which was equal to 239.1% and the number of Internet users reached 2.1 billion. These numbers highlight the required efforts for retrieving and classifying documents from a specific subject. A study by Autonomy [2], a Hewlett-Packard Company, mentions that the amount of unstructured data corresponds to 90% of the entire information set.

It is interesting to notice that the actual computational scenario, where studies on mobile configurations and lightweight protocols are growing [3][4], it is expected that the amount of data could increase in an unprecedented trend. This argument shows the importance of new retrieval techniques [5] and other experiences as key elements to provide some assessment to organizations.

As observed by Yan and Pederson [6], the efficiency of data retrieval is related on how the document database is configured. Thus, if a document exists within a large volume of data, inside a pre-classified set of data, it will probably be more efficient to gather specific information. In contrast, a natural assessment of large amounts of data would be difficult for humans due to the large amount of information available. Therefore, computational environments are more appropriate to process massive information sets. As Russel and Norvig [7] point out, these types of natural language processing are usually being executed through automatic document classification (ADC).

As mentioned in Baeza and Yates [8], the area of Information Retrieval (IR) has been the goal of many studies seeking automatic techniques capable to search and organize text documents using natural languages, including unstructured data. These approaches aim to gather, from user query, differentiated information available in libraries, storage systems, and specific places in order to yield dissemination of information.

This paper is characterized by an empirical case study, where the main objective is to present one assessment related to performance during the text pre-processing phase to build term bases from diffuse automatic document classification (ADC). Therefore, in this research we show a case-comparison between serial and parallel approaches. The serial element is implemented by adopting the RapidMiner software tool [9], which is recognized as the leading open-source system for data mining. On the other hand, the parallel side is represented by the MapReduce programming model (2013), utilized in the Hadoop [10] software environment.

Therefore, this paper is thus organized: theoretical aspects of RapidMiner, MapReduce, Hadoop, and HDFS are presented in Section 2. Section 3 presents the proposal and environment related to the present investigation. Empirical results based on experiments are presented and discussed in Section 4. Finally, in Section 5, conclusion and suggestions for further studies are presented.

2. Theoretical Aspects

In this section, we shall explore some conceptual aspects related to RapidMiner, MapReduce, Hadoop, and HDFS technologies. These four techniques are key elements for the proposal of this research work.

2.1. RapidMiner

RapidMiner [9] is an open-source system for data mining. It is available as a stand-alone application for data analysis and as a data mining engine for integration into other products. Furthermore, thousands of installations of RapidMiner in more than 40 countries give these users with competitive advantage [9].

Some essential characteristics of the RapidMiner [9] are:

- Data integration, Analytical ETL, Data Analysis, and Reporting in one single suite;
- Powerful but intuitive graphic user interface for the design of analysis processes;
- Repositories for process, data, and meta data handling;
- Single solution with meta data transformation: no trial and error, inspecting results directly during design stage;
- Single solution with on-the-fly error recognition support and quick fixes;
- Complete and flexible, with hundreds of data loading, data transformation, data modeling, and data visualization methods.

RapidMiner [9] is also well-known for some other operational features such as:

- Freely available open-source data mining and analysis system;
- Executes on every major platform and operating system;
- Most intuitive process design;
- Multi-layered data view concept which ensures efficient data handling;

- GUI mode, server mode (command line), or access via Java API;
- Simple extension mechanism;
- Powerful high-dimensional plotting facilities;
- Solution available for more than 500 operators for data integration and transformation, data mining, evaluation, and visualization;
- Automatic meta optimization schemes;
- Definition of re-usable building blocks;
- Standardized XML interchange format for processes;
- Graphic process design for standard tasks, scripting language for arbitrary operations;
- Machine learning library Weka [11] fully integrated;
- Access to data sources such as Excel, Access, Oracle, IBM DB2, Microsoft SQL, Sybase, Ingres, MySQL, Postgres, SPSS, dBase and Text files;
- Comprehensive data mining solution with respect to data integration, transformation, and modeling methods.

According to the RapidMiner [9] manual, it succeeds in nearly all cases in automatically determining the correct execution order of operators. In order to do so, RapidMiner uses information connection and an operator, the result of which is to be used by another operator, must obviously be executed before it. This works particularly well for this study as it is a sequential approach, since when the process was created its operators were properly lined up one after the other. There are other versions of RapidMiner software that implements parallel processes; however, in this research the free RapidMiner version has been used, with access to serial approaches only.

2.2. MapReduce

The actual world scenario causes the majority of organizations to deal with a requirement to process an enormous variety of data, of different types. Such processing is characterized by an information extraction operation from these data and aggregation of knowledge to the organization. The importance of such processing can be observed through two major aspects: the first, more evident, is the high volume of data; the second is related to the type of data, that is unstructured.

MapReduce is a programming model created by Google, typically used for processing large data sets with distributed configurations. The basic idea is to map and then reduce large tables, similarly to usual functional programming. However, it is different from a function programming language, since in the end MapReduce does not have the same original forms. The strength of the MapReduce programing model leads to the existence of several libraries that have been written in many programming languages.

As Lammal [12] states, the programming model of Google MapReduce is designed to process a large volume of data in a parallel fashion. Data processed by MapReduce are typically unstructured, such as texts and images. The parallel characteristic of this model allows a solution for any problem, dividing these into data matrixes which are executed without necessarily communicating among parallel tasks. In addition, as pointed out by Lammal [12], the programming model is based on simple concepts: input interaction; calculus for the peer key/value for each input; gathering of all intermediate values through a key use; interaction among resultant groups; reduction of each group.

As Google Code University [13] states, MapReduce was developed inside Google as a mechanism to process large amounts of data, in which only a distributed high performance system could manage to process such volume of data within reasonable elapsed time. In other words, this means utilizing a parallel paradigm. The approach provides an abstraction that allows for simple calculus of several problems, hiding details related to parallelization, data distribution, load balancing, and fault tolerance.

An open-source example of MapReduce is the implementation of Apache called Hadoop [10]. Hadoop utilizes a file system named Hadoop Distributed File System (HDFS) to store unstructured data. The Hadoop implementation allows applications execution in high-performance distributed configurations, with thousands of nodes and fault tolerance facilities.

Journal of Physics: Conference Series **540** (2014) 012001 doi:10.1088/1742-6596/540/1/012001

2.3. Hadoop

The Hadoop is a project from the Apache [10] characterized as an open-source framework implemented in Java language, which provides facilities for the development of distributed applications. Characteristics of these distributed applications are the requirement of large amount of data (reaching sometimes petabytes) and thousands of computer nodes.

The Hadoop project is based upon the Google File System (GFS) [14] and MapReduce. In 2006, the developer of Hadoop joined the Yahoo team and from this date the project is one entire system that works from the Web. In 2008, the project became an independent effort inside Apache and has been developed by a large community of programmers. The success of Hadoop can be measured by the number of large companies that are adopting the solution; some examples are: Facebook [15], Last.fm [16], Twitter [17], IBM [18], and Microsoft [19]. In addition, as an open-source effort in the official page of the project [10], it is possible to find JAR files necessary for Hadoop startup. In the Hadoop website, source codes, documentation, and several contributions can also be found.

2.4. Hadoop Distributed File System (HDFS)

Nowadays, distributed file systems are one of the most important research topics inside the area of distributed systems. Examples of these efforts are Google File System [14], Amazon Simple Storage Service [20], Ceph [21], Fraunhofer File System [22], Lustre [23], IBM General Parallel File System [24], and the Parallel Virtual File System [25].

The Hadoop Distributed File System [26] is a scalable and portable distributed file system written in Java for the Hadoop framework. HDFS is based on the GFS and has master/slave style architecture. The TCP/IP [4] stack protocol supports HDFS, in which a client uses the ClientProtocol with the NameNode through a port. DataNodes adopt the DataNodeProtocol with the NameNode, and these protocols execute a Remote Procedure Call (RPC).

An HDFS cluster comprises a unique NameNode, a master-server which is responsible for the system file management and controls access to all client files. On the other hand, there is a set of DataNodes, usually one per cluster node, which acts as manager for local storage. The NameNode operates all commands in the file systems, such as open, close, and rename (file or directory). Figure 1 below presents a typical HDFS architecture.

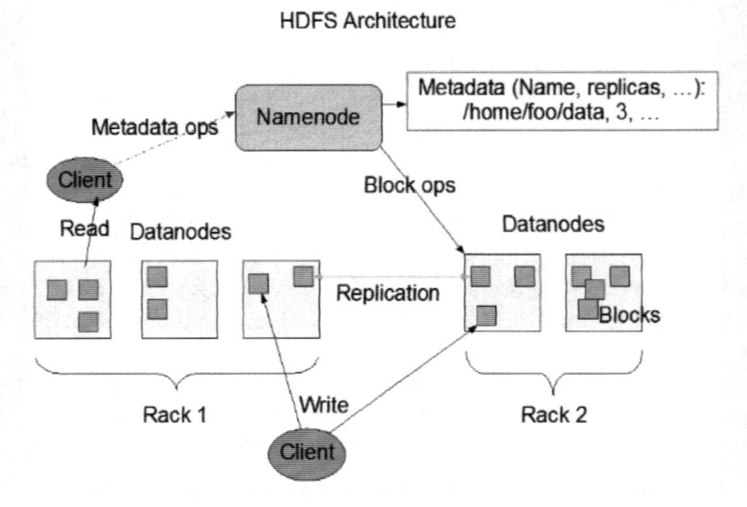

Figure 1. HDFS Architecture (Apache Hadoop, 2013).

HDFS exposes a file system namespace and allows user data to be stored in files. Internally, a file is split into one or more blocks and these blocks are stored in a set of DataNodes. A typical block size

Journal of Physics: Conference Series **540** (2014) 012001 doi:10.1088/1742-6596/540/1/012001

used by HDFS is 64 MB large. Reliability of the file system is reached through replication of data in several other nodes. Thus, it avoids the requirement of RAID storage.

HDFS does not provide high availability, because HDFS configuration requires one server (NameNode), thus this node represents a unique point that can fail. In case of failure, when server startup occurs it will be necessary to process all pending operations. This restart procedure can reach up to 30 minutes for large cluster environments. This period could be shorter in case a secondary NameNode has a procedure to gather regularly information in small time intervals about the NameNode master. As a result, the NameNode server is not obligated to execute all pending actions.

3. Proposal and test bed environment

The massive increase in data volume created with the intensive utilization of Web applications creates a new computational scenario where any conventional analysis of large-scale datasets has become outmoded.

In this section, we present the idea behind our study, which is to acquire knowledge on data processing by using an open-source system and then to consider a speedup technology for this process. Later, we have applies this assessment for an empirical process of text pre-processing to build term bases from a diffuse automatic document classification (ADC).

In order to reach our goal, we show a case-comparison study between two software environments, with different paradigms: one serial and one parallel. The serial software package is represented by the RapidMiner software tool [9], and the parallel paradigm is represented by the MapReduce programming model [27] utilized in the Hadoop [10] software environment.

Goller, G., J. Loning, T. Will, W. Wolff [28] can distinguish two stages in automatic document classification: the learning stage and the subsequent classification stage. In the learning stage, users define categories in which they are interested by giving sample documents (training examples) for each of these categories. Furthermore, the process of document classification involves three different steps: text preparation, characteristics selection, and later category definition.

Text preparation in the ADC approach is a relevant and costly step. This arguments is based on the required transformation process of the document set, considering a natural language, in a useful list of terms and also take into account a compatible form to prepare to extract knowledge.

In this research, as a test bed environment, we have first utilized resources from RapidMiner [9]. Figure 2 below shows a layout related to the text processing procedure, illustrating the interface feature from the environment, (claimed by RapidMiner developers), where they state that the software tool as a powerful and intuitive graphic user interface for the design of analysis process.

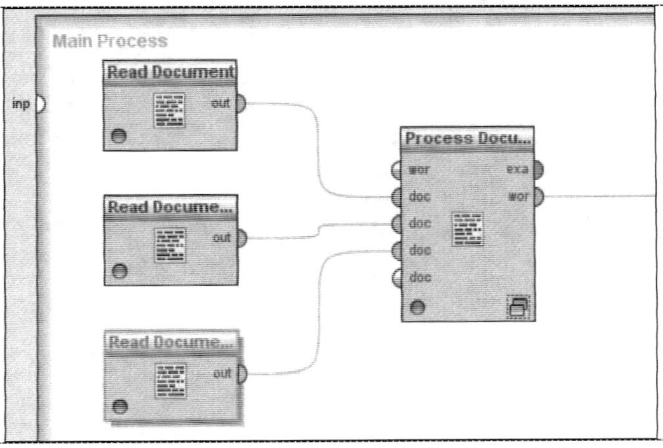

Figure 2. Example of text reading and processing.

On the other hand, Figure 3 below presents details of the text processing procedure illustrated in Figure 2 above. All texts from the same category are required to go through this pre-processing step in order to build a database which will provide storage facility for the most frequent terms and its respective grade of relevance for the category.

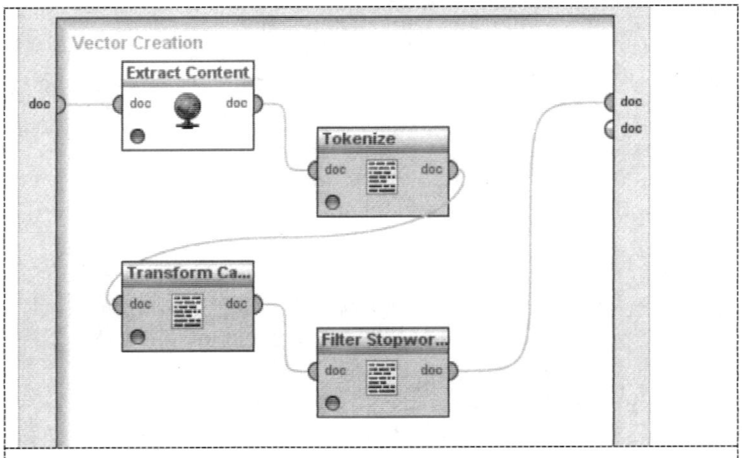

Figure 3. Detailing of document processing.

Figure 3 indicates that document processing is characterized first by the Extract Content operator, which extracts textual contents from a HTML document and yields text blocks extracted from the document. In sequence, the Tokenize operator divides text from a document in a token sequence. The next operation, Transform Case, transforms all characters from a document to uppercase or lowercase. Finally, the Filter Stopwords (Dictionary) operator extracts all words that are inside a stop-words list. This list is uploaded from a specific file inside the same operator. The stop-words list contains all words that are not relevant for the count of frequent terms inside the text document. In other words, this means that those words do not provide any influence for the definition of a text category.

In the second moment of the present study, targeted to build a database of a category with differentiated performance, we have considered the use of the MapReduce technique from the Hadoop software environment. In other words, this test bed configuration represents a modern option for the execution of one calculus when a large volume of Web text documents is necessary to build a category database.

The test bed environment for this stage of the study was designed while considering diverse situations. In the first scenario, the computational architecture was implemented with only one node working inside one core. The second configuration was characterized by two virtual machines (VMs), executing the Hadoop software environment in two different machines, with a unique core. The latter scenario represents an interesting distributed environment with one master and one slave node. Figure 4 below presents the main interface from the master VM executing Hadoop.

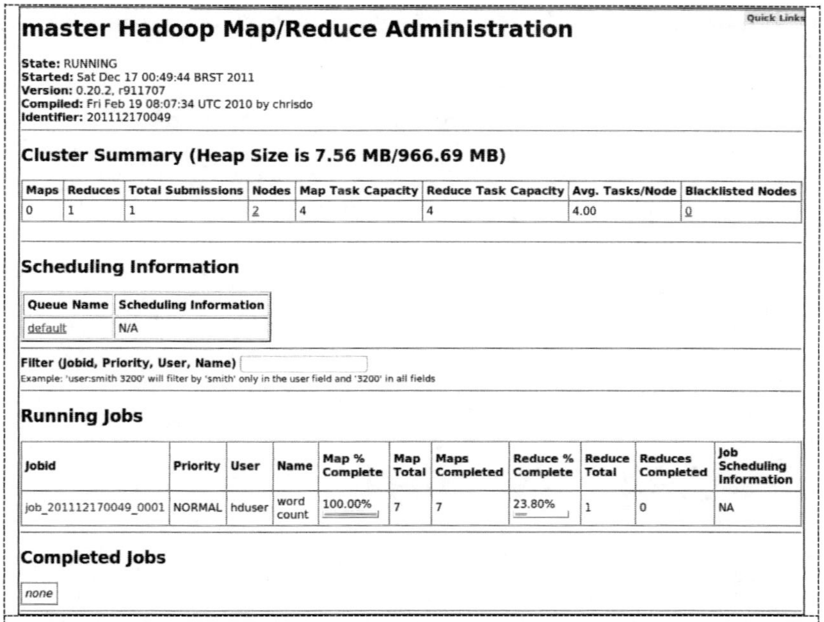

Figure 4. Hadoop interface at master VM node.

The Hadoop installation requires the Java Runtime Environment (JRE) and the Secure Shell (SSH) is also necessary. This package is used for initialization and finalization of script patterns created between nodes from the cluster. Compatible file systems with Hadoop inside the VMs inform slave node locations to the master node. Therefore, it will be possible to perform all main operations and task management of the environment.

As an ordinary procedure, the system performs executions of all local feasible operations. In other words, the priority is to execute any local task, where data exists locally. In the case of local data absence, the system will search for other configurations for the execution. The HDFS file system adopts a different data distributed storage policy to enhance environment performance.

One typical small Hadoop configuration, similar to the conceived test bed, comprises only one master node and several other slave nodes. In contrast, in a large Hadoop environment, the HDFS is managed through a unique and dedicated node that stores the NameNode and all file system indexes. In this environment, a secondary NameNode exists, which can mirror the structure execution of the master node, in case of failure from the master.

4. Empirical Results

In this section, results from our experiments utilizing the RapidMiner software package and the Hadoop/HDFS environment are presented. The objective is to have better assessment of the empirical process of text pre-processing phases to build term basis from a diffuse automatic document classification (ADC).

In the first configuration of our tests, the utilization of RapidMiner was considered, characterized by serial operations in an environment with one core. A set of four experiments were performed in this configuration, different among each due to text file sizes. In the first experiment, the operation was to process three different text files. These documents had sizes of 0.658MBytes, 1.356MBytes, and 1.56 MBytes, respectively. In the following test, the same text documents were processed. In the third execution, file sizes have been changed to 1.356 MBytes, 1.56 MBytes, and 13.6 MBytes, respectively. In the final experiment from this first set with the RapidMiner software package, the file sizes were changed once more to 1.56 Mbytes, 10.7 Mbytes, and 13.6 Mbytes. Figure 5 below shows results from these four initial experiments. Different text documents, with different sizes, were submitted for execution.

In the research presented by Alves et al. [29] several aspects which could influence the success of a Web application are presented. However, response latency is by far the most important element to any user of any software tool. This point can be observed in Figure 5, taking into consideration the issues of size and elapsed time for the respective responses. In other words, all tasks were executed successfully in an acceptable time threshold by the RapidMiner software.

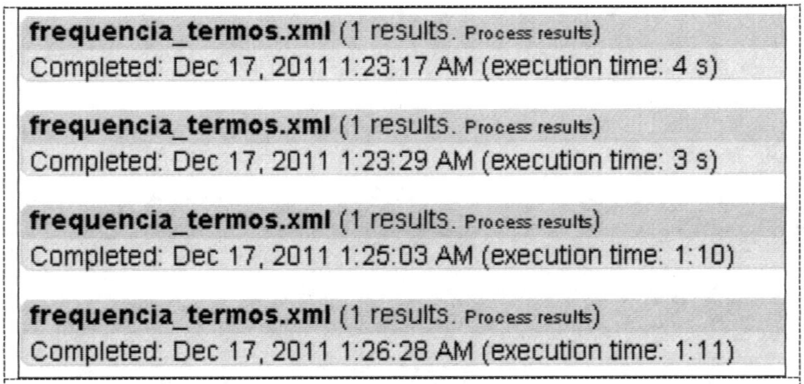

Figure 5. RapidMiner execution of four tests.

The second batch of experiments was to execute some case studies with the Hadoop environment. The first test considered, similarly to the RapidMiner experiment, one node acting as a master/slave, without data replication. In this test, documents had sizes of 0.658MBytes, 1.356MBytes, and 1.56 MBytes, respectively. The elapsed time to execute this task was 1 minute and 46 seconds, as illustrated in Figure 6 below.

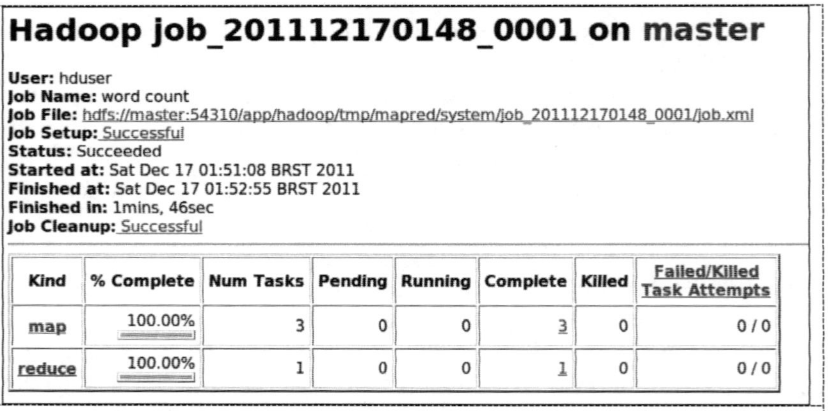

Figure 6. Hadoop test execution with one node.

In the second set of tests two different nodes were configured, where the same text documents (sized 0.658MBytes, 1.356MBytes, and 1.56 Mbytes) were processed. The resulting elapsed time was 32 seconds, as shown in Figure 7 below. The application failed to process other experiments similar to those performed by RapidMiner, with different files and different file sizes. This problem seems to be a hardware limitation where some nodes do not have enough memory and therefore have crashed.

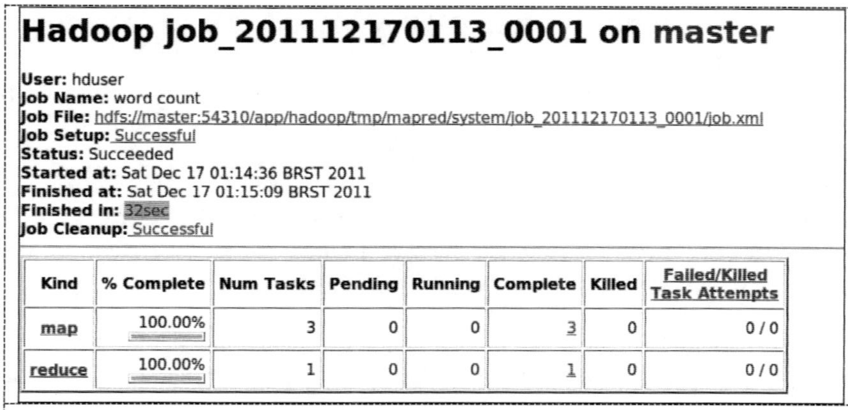

Hadoop job_201112170113_0001 on master

User: hduser
Job Name: word count
Job File: hdfs://master:54310/app/hadoop/tmp/mapred/system/job_201112170113_0001/job.xml
Job Setup: Successful
Status: Succeeded
Started at: Sat Dec 17 01:14:36 BRST 2011
Finished at: Sat Dec 17 01:15:09 BRST 2011
Finished in: 32sec
Job Cleanup: Successful

Kind	% Complete	Num Tasks	Pending	Running	Complete	Killed	Failed/Killed Task Attempts
map	100.00%	3	0	0	3	0	0 / 0
reduce	100.00%	1	0	0	1	0	0 / 0

Figure 7. Hadoop test execution with two nodes.

5. Conclusion and Suggestions for Further Research

The goal of this case-comparison study is to exploit differences between serial and parallel paradigms, especially when a large volumes of data from Web documents is utilized to build a category database. In the literature, several studies point out that the distributed document processing of unstructured data have presented interesting levels of efficiency.

The tested database stores normalized terms within their category in order to assess their level of importance. This alternative proposal of Hadoop implementation was built for an Automatic Text Document Categorization System (ATDCS) based on two concepts: similarity and accuracy. These two concepts are input to a fuzzy modeling which ultimately allows specifying the level of relevance of a text to a specific category.

Preliminary results in the test environment did not present favorable performance increment for the distributed implementation if the volume of unstructured data is on the order of 10^6 (Mega Bytes). Still, it is estimated that Hadoop performs better when the amount of data is on the order of 10^15 (Peta Bytes). Normally there is the expectation that in parallel programs processing should be more efficient, since the process is split into parts, where each can be executed simultaneously. In cases where the volume of data is not significant, the MapReduce solution implemented by Hadoop is not more interesting than the serialized solution from RapidMiner.

References

[1] Royal Pingdom. [online] http://royal.pingdom.com/2012/01/17/internet-2011-in-numbers/ (Accessed 15 March 2013).

[2] Autonomy: an HP company. What is big data? [online] http://www.autonomy.com/content/Technology/what-is-big-data/index.en.html (Accessed 15 March 2013).

[3] F.J. Knaesel, M.A.R. Dantas. High-performance sharing of consistent data in ad hoc networks. *Int. J. of Networking and Virtual Organisations*, 2009 Vol.6, No.3, pp.259 – 269.

[4] Diogo R. Viegas, M.A.R. Dantas, Michael A. Bauer. A case study of transport protocols to improve the execution of applications in virtual organisations utilising multicluster network configurations. *Int. J. of Networking and Virtual Organisations*, 2010 Vol.7, No.5, pp.433 – 451.

[5] S. Sendhilkumar, G.S. Mahalakshmi, Context-based citation retrieval, *Int. J. of Networking and Virtual Organisations*. 2011 Vol.8, No.1/2, pp.98 – 122.

[6] Yang, Y; Pederson, J, 1997: A Comparative Study on Feature Selection in Text Categorization. *In Proceedings of 14th International Conference on Machine Learning*, Morgan Kaufmann Publishers, San Francisco, US (1997), 412-420.

[7] Russell, Stuart Jonathan, Norvig, Peter. 2009. Artificial intelligence: a modern approach. 3rd edition. Prentice-Hall
[8] Baeza-Yates, Ricardo, Ribeiro-Neto, Berthier. (1999) Modern Information Retrieval. 1 ed., United Kingdom, Addison-Wesley.
[9] RapidMiner. [online] http://rapid-i.com (Accessed 15 March 2013).
[10] Apache Hadoop. [online] http://hadoop.apache.org/ (Accessed 15 March 2013).
[11] Weka. [online] http://www.cs.waikato.ac.nz/ml/weka/index.html (Accessed 15 March 2013).
[12] Lammal, Ralf. Googles MapReduce Programming Model Revisited (2007). [online] http://logic.cse.unt.edu/tarau/teaching/SoftDev/docs/mapReduce.pdf (Accessed 15 March 2013).
[13] Google Code University. Introduction to Parallel Programming and MapReduce. [online] http://code.google.com/intl/pt-BR/edu/parallel/mapreduce-tutorial.html (Accessed 15 March 2013).
[14] Google File System - GFS. [online] http://research.google.com/archive/gfs.html (Accessed 15 March 2013).
[15] Facebook. [online] http://engineering.twitter.com/2010/04/hadoop-at-twitter.html (Accessed 15 March 2013).
[16] Last.fm. [online] http://www.lastfm.com.br/ (Accessed 15 March 2013).
[17] Twitter. https://twitter.com/ (Accessed 15 March 2013).
[18] International Business Machines (IBM). [online] http://www-01.ibm.com/software/data/infosphere/hadoop/ (Accessed 15 March 2013).
[19] Microsoft. [online] http://www.microsoft.com/en-us/sqlserver/solutions-technologies/business-intelligence/big-data.aspx (Accessed 15 March 2013).
[20] S3. [online] http://aws.amazon.com/pt/s3/ (Accessed 15 March 2013).
[21] Ceph. [online] http://ceph.com/ceph-storage/file-system (Accessed 15 March 2013).
[22] FhGFS. [online] http://www.fhgfs.com/cms/ (Accessed 15 March 2013).
[23] Lustre. [online] http://wiki.lustre.org/index.php/Main_Page (Accessed 15 March 2013).
[24] GPFS. [online] http://www-03.ibm.com/systems/software/gpfs/ (Accessed 15 March 2013).
[25] PVFS. [online] http://www.pvfs.org/ (Accessed 15 March 2013).
[26] HDFS Architecture Guide. Apache Software Foundation. [online] http://hadoop.apache.org/common/docs/current/hdfs_design.html (Accessed 15 March 2013).
[27] MapReduce. [online] http://research.google.com/archive/mapreduce.html (Accessed 15 March 2013).
[28] Goller, G., J. Loning, T. Will, W. Wolff. Automatic Document Classification: A thorough Evaluation of Various Methods. *Internationalen Symposiums für Informationswissenschaft*, Nov. 2000, pp. 145-162.
[29] Alves, V., Dantas, M.A.R., An Approach Based on Speedup Web Paradigm, submitted to Webmedia 2013.

Journal of Physics: Conference Series **540** (2014) 012002 doi:10.1088/1742-6596/540/1/012002

Towards a Resource Reservation Approach for an Opportunistic Computing Environment

Eliza Gomes, M.A.R. Dantas

Research Laboratory of Distributed Systems (LaPeSD)
Department of Informatics and Statistic (INE)
Federal University of Santa Catarina (UFSC)
88040-900 - Florianopolis,SC - Brazil

E-mail: eliza.gomes@posgrad.ufsc.br, mario.dantas@ufsc.br

Abstract.
Advanced reservation has been used in grid environments to provide quality of service (QoS) and to guarantee resources available at the execution time. However, in grid subtypes, such as opportunistic grid computing, it is a challenge provides QoS and guarantee of availability resources. In this article, we propose a new advanced reservation approach which offers to a user the possibility to select resources in advance for a future utilization. Therefore, the main goal of this proposal is to offer a best effort feature to a user from an opportunistic configuration. In these types of environments, it is not possible to provide QoS, because, usually, there are no guarantees of resources availability and, consequently, the execution of users applications. In addition, this research work provides a way to organize executions, what it can improve the scheduling and system operations. Experimental results, carried out through a case study, shown the efficiency and relevance of our proposal.

1. Introduction

In recent years we have seen increasing use of supercomputers to execute applications which require high performance computing [1]. On the other hand, personal computers have increasingly computational power to, often, execute simple applications. In the face of this scenario, to cluster these devices, as opportunistic configurations, has become a widely paradigm used (and required) by industrial and academic communities. Opportunistic grid configurations are a sub-type of environment, which utilize idle computational resources from personal machines, companies and laboratories, to form, dynamically, a computational grid [2]. Examples of middleware for opportunistic grid are the projects: SETI@home [3], BOINC [4] and InteGrade [5].

Opportunistic systems can be characterized as meta-schedulers, because they receive requests from users and scheduling their applications to different local schedulers. However, these execution requests of tasks are, usually, randomly and disorderly. In other words, requests may not exist for long periods and then a system receives several requests. This scenario illustrates that in any system can become hard to process users' requests.

This article presents a proposal of an implementation of an advanced reservation of resources to opportunistic grid environments. The advanced reservation concept can be understood as the ability to user to reserves resource in advanced to be used at specific time in the future

[6]. Resources commonly reserved are: CPU and amount of memory, disk space and links [7]. The InteGrade project was utilized as a middleware to opportunistic grid, where we added our contribution.

This proposal comprises an implementation of an additional module in the InteGrade architecture which enables users to request resources in advance. The main goal is to offer to user a best effort, once is not possible to guarantee QoS, due to dynamism of the environment. In addition, our proposal aims to improve the applications scheduling, in looking for an advanced reservation slot, somehow, organizes executions, once the applications will be organized in queue, waiting for the beginning of the execution.

The proposal was evaluated through a case study which simulates the module behavior. Experimental results show the efficiency of the mechanism and a best effort offered to user.

This article is organized as follows: section 2 provides a review related to the concepts of co-scheduling, co-allocation and co-reservation to build a meta-scheduler while section 3 presents related works. Section 4 describes the proposed architecture. In section 5 the implementation is simulated through case study. Section 6 concludes the article and suggests some future works.

2. Co-Scheduling, Co-Reservation and Co-Allocation

Meta-schedulers are characterized by receiving requests from users and schedule applications to local cluster scheduler. They also have the function of orchestrate multi-cluster configurations [8]. The use of advance reservation in multi-schedulers requires efficient algorithms to co-scheduling, co-reservation and co-allocation.

2.1. Co-Scheduling

Scheduling determines where and when jobs are executed and how many resources will be allocated. Co-scheduling can be understood as the ability to schedule jobs to execute at the same time at different processing nodes [8]. There are two types of co-scheduling [9], Gang and Loosely.

- Gang: schedule process of a job at the same time at different allocated nodes or processors to this job.
- Loosely: originated from NOW (Network of Workstations) environments, seeks to solve the problems to scheduling serial and parallel workload together. This approach is based on the local scheduling and on communication organization of the involved nodes.

2.2. Co-Reservation

The co-reservation function is characterized by the coordination use of cluster resources in sequence as a grid configuration [8]. Usually are adopted timeslot tables to show the percentage of available resources allocated during a determinate period of time. These tables help to control current allocations and future reserves [10]. Figure 1 shows an example of timeslot table with As representing current allocations and Rs as future reservations.

2.3. Co-Allocation

Co-allocation can be defined as the simultaneous use of grid resources through multi-clusters [8]. According to Netto and Buyya [7], co-allocation is the resource allocation process from multiple administrative domains in a coordinated.

3. Related Research Work

In this section, we present some related works that present some related characteristic related to our research.

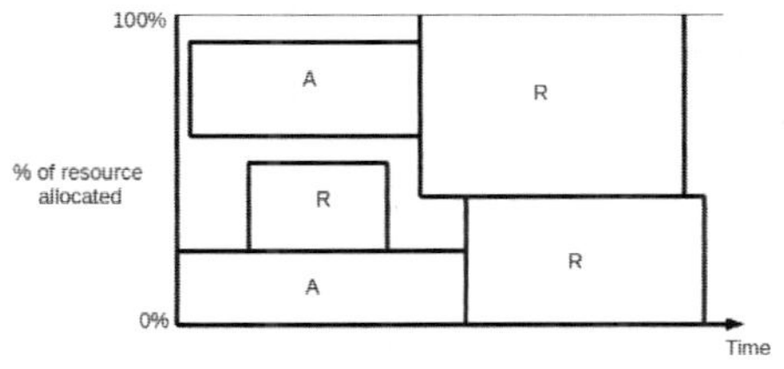

Figure 1. Timeslot Table

Advance reservation of resources method presented at Ferreira and Dantas [11] research work enables user to reserve resources between different clusters in a multi-clusters environment. The proposal uses semantic integration of multiple ontologies to match users requests with description of the system configurations. Furthermore, it includes fuzzy logic to control the distribution of tasks among clusters, since applications are subdivided into tasks and it can be executed in several clusters.

Sulistio et. al. [12] present a data structure based on Segment Tree and Calendar Queue, to provide efficiently management of advance reservations. This data structures, named as GarQ (Grid advanced reservation Queue), found available nodes, add requests and delete existing reservations efficiently and quickly, as shown by the experimental results. Furthermore, The authors have developed a mechanism to search for free time intervals closest to the requested reservation, if it was reject before.

Melo Martins et. al. [13] research presents a management mechanism design to opportunistic grid computing environments to handling the execution of applications with time deadlines set by users during their submission in the system. This mechanism is based in dynamic scheduling and re-scheduling and it was developed to execute in InteGrade middleware. In additional, the research proposal uses the advanced reservation mechanism to reserves the choose node to execute the submitted application if the node is busy, in other words, executing other application. As experimental results the authors carried out simulations in different opportunistic computing scenarios.

4. Proposal

In this section, it is presented our proposed architecture which provides an advanced reservation for opportunistic environments. Since grid environments are dynamic, an advanced reservation feature becomes important aspect, because it improves the scheduling process from an application [14], once the scheduler will know in advance when and, perhaps, where an application will be executed and an availability of resources for applications.

Researches such as [15] and [16]have the objective to reach the quality of service (QoS) in grid environments using advanced reservation. The main goal of these works is to guarantee that the resources will be available at some time in the future. In contrast to these proposals, our research work aims to achieve the best possible quality of service, once we are developing a proposal for an opportunistic environment. In other words, our proposal does not guarantee the availability and, therefore the application execution. Nevertheless, our proposal is relevant since we seek to improve both usability of the system to a user, as to organization and operation of the system.

The InteGrade project [5] was employed to develop our proposal. InteGrade is an opportunistic grid computing middleware whose purpose is to use idle computing capacity of workstations, academic laboratories and personal computers to execute applications which require high performance computing [17]. Its main goal is to offer to grid users high performance computing without degrading the performance of their machines, always prioritizing the

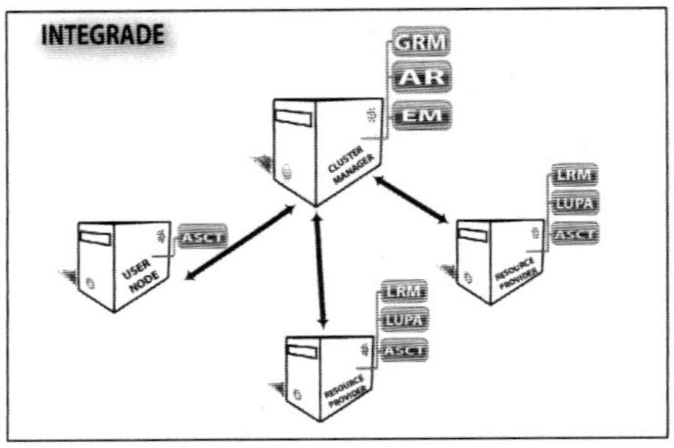

Figure 2. InteGrade Architecture

execution of local applications. The InteGrade was chosen for the development of this research because no advanced reservation exists in its basic architecture, it is a project developed through the partnership between Brazilian universities. Furthermore, it has open source and easy access, and the availability of efficient support.

4.1. InteGrade

Currently, InteGrade executes three types of applications [18]:

- *Sequential Applications:* single applications, with only one executable binary which require only one node to execute.
- *Parametric or Bag-of-Tasks Applications:* several application copies are distributed to different cluster's nodes. Each copy processes different sets of parameters and input files independently of the data of the others nodes.
- *Parallel Applications:* parallel applications in the InteGrade are based in BSP and MPI models.

The basic architecture of InteGrade grid is a cluster, in other words, a set of machines connected to a local network and organized hierarchically [18]. Figure 2 shows the major components of the InteGrade cluster. Each cluster has a Cluster Manager node which is responsible for the application scheduling and the storage information of each node belonging to the cluster. It has a set of nodes which offers computing resources to the cluster, named Resources Provider nodes. Furthermore, the cluster has User Nodes in which user can submit application to execution without provide their resources to the cluster.

The InteGrade architecture is composed of modules which they are responsible for cluster operation. The modules, depicted in Figure 2, are:

- *ASCT - Application Submission and Control Tool:* it is a friendly graphical users interface which enable user to submit applications to the cluster [17]. Users can specify hardware requirements and software platform, and monitor their application execution.
- *LRM - Local Resource Manager:* it is executed each cluster node. It collects information about the node status, such as memory, disk, CPU and network utilization. This information is send periodically to the GRM. It is also responsible for instantiating and executing applications schedule to its node.
- *GRM - Global Resource Manager:* it is a manager cluster. GRM receives information collected by LRMs about quantity and availability of resources belonging to the cluster, and it uses them to scheduling applications. It is responsible to communication with GRMs of others clusters.

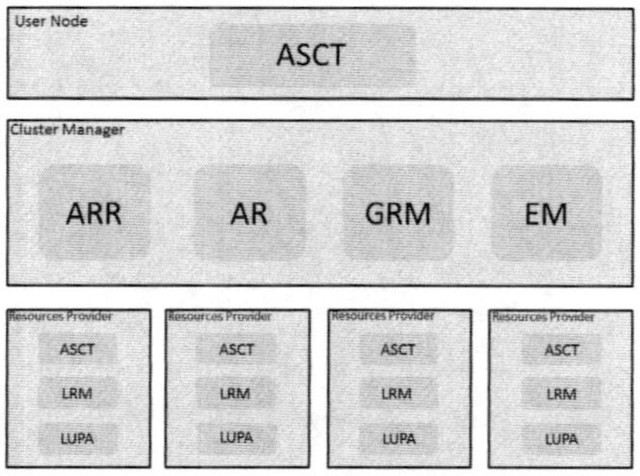

Figure 3. ARR Module

- *AR - Application Repository:* storage applications which are submitted by the ASCT to be executed later in the cluster.
- *EM - Execution Manager:* maintains information about applications submitted to cluster, such as submission and execution state, input and output parameters [18].
- *LUPA - Local Usage Pattern Analyzer:* it executes each cluster node and establishes usage pattern through analysis of resources use [19]. These patterns are used to predict idle period of the node and improve the applications scheduling.

4.2. Proposed Architecture

To enhance the InteGrade architecture, we proposed an advanced reservation of resources (ARR) module. This module enables users to require a reservation in advance resources from a cluster to be used in the future. The Figure 3 shows the insertion of the ARR module in the Cluster Manager in order to maintain the basic structure of InteGrade.

The ARR is composed of service modules which are responsible for its operation. These modules are: Reservation Service, Execution Service and Notification Service.

The Reservation Service deals with the management of reservation request. In other words, it verifies if there is a free time interval and reallocates the reservation if an interval resquested is not available. To reach this goal, the service receives from User Node information necessary for executing an application, originally provided by InteGrade. Examples are: application type (sequential, parametric or parallel), quantity for resources to application execution, identification of application in the Application Repository and the information relating to advanced reservation and stored them in the database.. The information required to advanced reservation is:

- *start-time*: execution start time;
- *end-time*: execution end time;
- *user*: user name to handle stored reservations;
- *email*: user e-mail to notifications.

The Execution Service is responsible for requesting the beginning of application execution. Then, the service searches the execution information in the database and sends it to Cluster Manager which starts the procedures to application execution.

The Notification Service warns users, by e-mail, when the process failure or in the initialization or completion of application execution.

The following section shows the application execution process in InteGrade using the Advanced Reservation of Resource module.

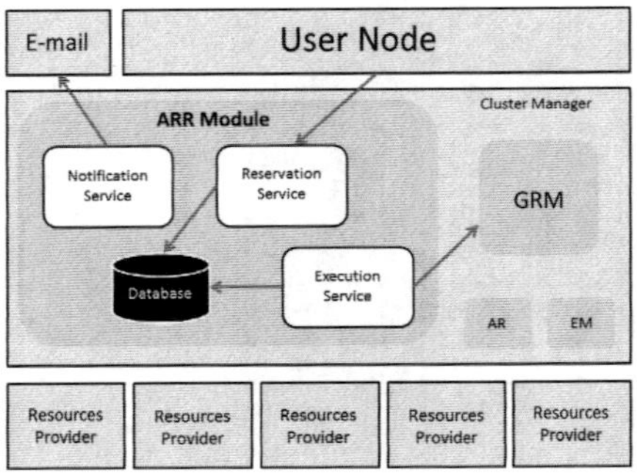

Figure 4. ARR Architecture

4.3. Execution Process

The execution process starts when user registers the application on AR. Then, user can choose between immediate and scheduled execution. Both for immediate execution and for scheduled execution, user specifies information such as quantity of resources to execution, application type (sequential, parametric or parallel) and input files. When user chooses scheduled execution, it is necessary, additionally, to specify information such as start execution time, end execution time and e-mail.

In the next step, user submits the execution scheduled request. At this moment, the ASCT sends information regarding the resource specifications to GRM. This module searches for nodes, in the cluster, which match with the resource requested by user. If the resources do not exist, user is notified by the ASCT. Otherwise, the ASCT sends the information regarding reservation to the Reservation Service. Then, the module verifies in the database whether there is, or not, reservation in the requested period. If there is not, the reservation is successful. Otherwise, the Reservation Service suggests an available period of time closest to the specific what it can be accepted or not. Upon the specification of period, the information regarding the resources, application, reservation and input files is stored in the database and the reservation is successful.

When executing the application, the Execution Service retrieves the information regarding the resources, in the database, and sends it to the GRM. The GRM verifies whether the node previously selected is still available to application execution. If the node still to belong to the cluster, though it is executed an application, the GRM saves the application status and verifies if there are in the cluster other available nodes to complete the execution. This occurs because we assume that a scheduled execution takes priority over immediate execution. In case of the node does not belong to the cluster, the GRM verifies whether there are other nodes with characteristics which match with user's requirements.

After selecting the node, the GRM sends an execution request to LRM of the candidate node. At this point, the LRM verifies, if indeed, the node has available resources. This procedure is necessary because the GRM maintains an approximate vision of the availability of nodes resources belonging to the cluster. If the node has not available resources, the LRM notifies the GRM which searches, again, for candidate node. However, if the candidate node is available, LRM requests the application to the Application Repository, it requests to Execution Service the input files and it executes the application. The LRM sends a notification of acceptance and after the completion of the execution to the Notification Service which notifies user through e-mail.

Table 1. Characteristics of machines

Machine	Quantity of Memory (MB)	Quantity of CPU
node 1	3072	1
node 2	4096	1
node 3	2048	1

Table 2. Reservations stored in the database

User	Start Time	End Time	CPU Usage (%)	Free Memory (MB)	Application Type	Reserved Node
A	09:00am	11:00am	< 70	≥ 2000	Parametric	node 2
B	12:00pm	02:00pm	< 50	≥ 1024	Sequential	node 3
C	06:00pm	07:00pm	without specifications	without specifications	Parallel	node 1
D	12:00pm	02:00pm	< 50	≥ 1024	Sequential	node 2

5. Experimental Results

In this section, we tested our proposal through of a case study which illustrates the InteGrade behaviour with the implementation of Advance Reservation of Resources module.

5.1. Environment

The cluster environment is composed of one Cluster Manager (GRM) and three Resources Provider (LRM). The Table 1 describes the quantity of memory of each LRM that belong to the cluster.

The GRM implements a heuristics to the alternation of the nodes which will meet the requisitions. The goal is to prioritize nodes which do not received requisitions recently, thereby the GRM minimizes the possibilities of the requisitions be sent to the same node [20]. The Table 2 shows the reservations stored in the database.

Although user to provide resources specifications in the reservation request, the reservation module reserves the selected node, instead of the resources. The goal of this method is to have no more than one reservation per node what would overload it.

5.2. Case Study

In this subsection we present a case study that simulates the system behavior with conflicts of the new reserves. The Table 3 presents four requests to be carried by four users.

The reservation request of user G is accepted without restrictions, once there is not reservation with the same period stored in the database. Then, in this case, the node 3 is reserved. The reservation request of user E is at the same time of user B, but it is accepted. This occurs because the environment is composed of three nodes there is still the node 1 available at this time. Then, the node 1 is reserved to user E.

On the other hand, reservation of user F is refused by the system because the period between 12:00pm and 02:00pm all nodes are reserved. Therefore there would be no sufficient resources to satisfy user F request. To solve this problem, the Reservation Service offers the reallocation of reservation to user. In this case, the module suggests the period between 03:00pm and 05:00pm, in other words, the free time interval closest to the one chosen by user. If user accepts the

Table 3. Reservations requested by users

User	Start Time	End Time	CPU Usage (%)	Free Memory (MB)	Application Type
G	08:00pm	09:00pm	< 50	> 2300	Parallel
E	12:00pm	02:00pm	< 70	> 1024	Sequential
F	12:00pm	02:00pm	< 70	> 1024	Parallel
A	10:00pm	11:00pm	< 70	> 1024	Parametric

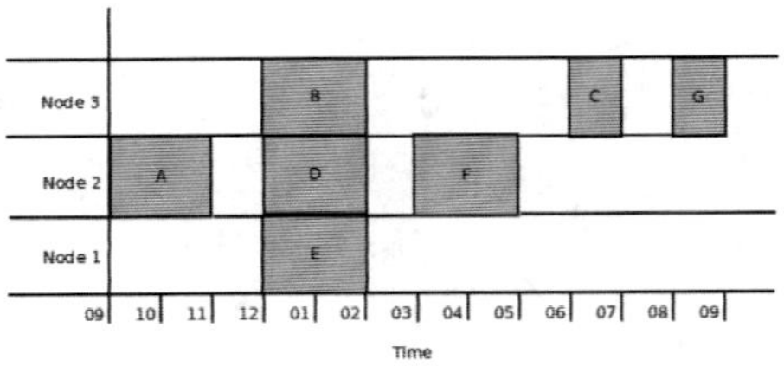

Figure 5. Timeslot table

suggestion, the node 2 will be reserved. The reservation request of user A is not accepted, since this user already has a reservation stored in the database. We adopt as a rule, that only one reservation can be stored in the database per user, this procedure avoids an only user reserves several periods of time and precludes the use of this mechanism by other users of the cluster.

As result, the Figure 5 shows the reservation allocation through timeslot table. The execution queue is sorted, firstly, by start time and then by request order.

6. Conclusions and Future Works

In this article, it was proposed an architecture which consisted in an approach of advance reservation of resources in opportunistic grid computing environments. The main goal was to improve applications scheduling and to provide a best-effort feature to users. The proposal was developed using the InteGrade project opportunistic grid computing middleware.

Therefore, a review from the InteGrade architecture was presented. In a second moment, it was presented the proposal and its operation with the InteGrade. It was, also, demonstrated an execution process of an application, when a user chose a scheduled execution. The approach was evaluated successfully through case study, where an InteGrade simulation behavior was executed in a real environment, where the combined configuration used an advanced reservation.

As a future work we intend to add in ARR module a service which manipulates the information collected by LUPA about usage patterns of each node and it generates idle periods prediction of entire the cluster to help users in the chosen the most appropriated period to execute their applications.

Acknowledgment

The authors would like to thank the attention of Professor Alfredo Goldman and his disposition to discuss part of the work presented in this article.

References

[1] Top500. http://www.top500.org/, January 2013.

[2] F. Kon and A. Goldman. Grid computing: Fundamental concepts and concrete cases. *Proceedings of the Tomasz Kowaltowski e Karin Breitman (Org.)*, pages 55–104, 2008.

[3] SETI@home. http://setiathome.ssl.berkeley.edu/, January 2013.

[4] BOINC. http://boinc.berkeley.edu/, January 2013.

[5] InteGrade. http://www.integrade.org.br, January 2013.

[6] W. Smith, I. Foster, and V. Taylor. Scheduling with advanced reservations. In *Parallel and Distributed Processing Symposium, 2000. IPDPS 2000. Proceedings. 14th International*, pages 127–132. IEEE, 2000.

[7] M.A.S. Netto and R. Buyya. Resource co-allocation in grid computing environments. *Handbook of Research on P2P and Grid Systems for Service-Oriented Computing: Models, Methodologies and Applications. IGI Global*, 2009.

[8] J. Qin, M.A.R. Dantas, and M. Bauer. Application programmers interactions enhancing a meta-scheduler in multi-clusters environment. Unpublished.

[9] A.C. Sodan. Loosely coordinated coscheduling in the context of other approaches for dynamic job scheduling: a survey. *Concurrency and Computation: Practice and Experience*, 17(15):1725–1781, 2005.

[10] I. Foster, C. Kesselman, C. Lee, B. Lindell, K. Nahrstedt, and A. Roy. A distributed resource management architecture that supports advance reservations and co-allocation. In *Quality of Service, 1999. IWQoS'99. 1999 Seventh International Workshop on*, pages 27–36. IEEE, 1999.

[11] DJ Ferreira, MAR Dantas, and M.A. Bauer. An ontological-fuzzy approach to advance reservation in multi-cluster grids. In *Journal of Physics: Conference Series*, volume 256, page 012021. IOP Publishing, 2010.

[12] A. Sulistio, U. Cibej, S.K. Prasad, and R. Buyya. Garq: An efficient scheduling data structure for advance reservations of grid resources. *International Journal of Parallel, Emergent and Distributed Systems*, 24(1):1–19, 2009.

[13] M.R. Melo Martins, B.T. Pereira Gomes, J.F. Gonçalves, and F.J. da Silva e Silva. Execution management of applications with runtime restrictions on opportunistic grids environments. In *PESARO 2012, The Second International Conference on Performance, Safety and Robustness in Complex Systems and Applications*, pages 1–7, 2012.

[14] L. Tomás, C. Carrión, B. Caminero, and A. Caminero. Meta-scheduling in advance using red-black trees in heterogeneous grids. In *Parallel & Distributed Processing, Workshops and Phd Forum (IPDPSW), 2010 IEEE International Symposium on*, pages 1–8. IEEE, 2010.

[15] M. Siddiqui, A. Villazón, and T. Fahringer. Grid capacity planning with negotiation-based advance reservation for optimized qos. In *Proceedings of the 2006 ACM/IEEE conference on Supercomputing*, page 103. ACM, 2006.

[16] A. Di Stefano, G. Morana, and D. Zito. Advanced reservation in grid using pbs. In *Workshop on Enabling Technologies: Infrastructure for Collaborative Enterprises, 2008. WETICE'08. IEEE 17th*, pages 216–221. IEEE, 2008.

[17] Andrei Goldchleger, Fabio Kon, Alfredo Goldman, Marcelo Finger, and Germano Capistrano Bezerra. Integrade: Object-oriented grid middleware leveraging idle computing power of desktop machines. *Concurrency and Computation: Practice and Experience*, 16(5):449–459, March 2004.

[18] F.J. da Silva e Silva, F. Kon, A. Goldman, M. Finger, R.Y. de Camargo, F.M. Costa, et al. Application execution management on the integrade opportunistic grid middleware. *Journal of Parallel and Distributed Computing*, 70(5):573–583, 2010.

[19] Marcelo Finger, Germano C. Bezerra, and Danilo M. R. Conde. Resource use pattern analysis for predicting resource availability in opportunistic grids. *Concurrency and Computation: Practice and Experience*, 2009.

[20] Andrei Goldchleger. Integrade: A middleware system to opportunistic grid computing. Master's thesis, University of Sao Paulo, December 2004.

Modeling Remote I/O versus Staging Tradeoff in Multi-Data Center Computing

Ibrahim H Suslu

Department of Computer Science, North American College, Houston, TX 77038

E-mail: ihsuslu@northamerican.edu

Abstract. In multi-data center computing, data to be processed is not always local to the computation. This is a major challenge especially for data-intensive Cloud computing applications, since large amount of data would need to be either moved the local sites (staging) or accessed remotely over the network (remote I/O). Cloud application developers generally chose between staging and remote I/O intuitively without making any scientific comparison specific to their application data access patterns since there is no generic model available that they can use. In this paper, we propose a generic model for the Cloud application developers which would help them to choose the most appropriate data access mechanism for their specific application workloads. We define the parameters that potentially affect the end-to-end performance of the multi-data center Cloud applications which need to access large datasets over the network. To test and validate our models, we implemented a series of synthetic benchmark applications to simulate the most common data access patterns encountered in Cloud applications. We show that our model provides promising results in different settings with different parameters, such as network bandwidth, server and client capabilities, and data access ratio.

1. Introduction

Data-intensive Cloud applications usually take place in two phases: data generation/collection and data analysis. In the data generation/collection phase, large amounts of data are generated by applications running on different resources or collected from remote instruments, such as sensors, detectors, social networks or other Web applications. In the data analysis phase, collected data is analyzed and another set of data may be generated. If in any of the data collection or analysis phases more than one data centers are involved, efficient data access mechanisms would be an important factor in the end-to-end performance of these applications.

There are mainly four different data access techniques Cloud application developers consider: *(i)* accessing data streams remotely over the network (remote I/O); *(ii)* moving application where data is located; *(iii)* moving data where application runs (staging); and *(iv)* moving both data and application to an intermediate location (hybrid model). Each data access technique has advantages and disadvantages based on the characteristics of the environment and applications. Therefore, based on application needs, an appropriate data access technique should be used.

In most Cloud applications, the datasets required by the application are either transferred to a temporary space close to the computation (staged-in) or accessed remotely over the network (remote I/O). Due to the nature of applications and the interconnects sites, there are many factors affecting end-to-end performance such as I/O access pattern of the application, size of dataset needed, and network characteristics. Although both data access techniques are widely used, the Cloud application developers generally choose one over the other intuitively without making any scientific comparison specific to their application since there is no generic model available that they can use. To the best of our knowledge, there has not been an extensive study comparing both data access techniques and providing clear guidelines of which method to use in which particular case.

In this paper, we propose a generic model for the Cloud application developers which would help them to choose the most appropriate data access mechanism for their specific application workloads. The current approach to find the best data access method is based on active learning. First, the application needs to be run in the same environment with all possible combinations. Once the combination that is the best fit for that environment is discovered, the correct data access technique can be found. Our model, however, provides the best data access technique before running the application.

We initiallly define the parameters that potentially affect the end-to-end performance of the multi-data center Cloud applications which need to access large datasets over the network. To test and validate our models, we implement a series of synthetic benchmark applications to simulate the most common data access patterns encountered in Cloud applications. We show that our model provides promising results in different settings with different parameters, such as network bandwidth, server and client capabilities, and data access ratio. Cloud application designers and developers can use our model to decide which data access technique is right for their application.

2. The Theoretical Model

Modeling the data access techniques in a Cloud computing environment is not a well-studied area. Staging the data before it is needed can be very complicated because of the dynamic nature of the Cloud environment such as multi-data center configuration, dynamic resource allocation, and the real-time network traffic between sites. Limitations of bandwidth, storage space at certain sites, and data specifications can affect the staging of the data.

In order to model the tradeoff between staging and remote I/O, we first define the parameters that can affect the end-to-end performance of the multi-data center Cloud applications. The application specific values for these parameters should be provided by the user. And developed model would need to evaluate these parameters not only individually, but also as a combination to find out which data access technique is the best for the specific application. The parameters of interest are listed below:

- Rl: time to read from local disk to local memory
- Rr: time to read from remote disk to remote memory
- Wl: time to write from local memory to local disk
- Wr: time to write from remote memory to remote disk
- Ns: time to send data over network via a staging protocol
- Nr: time to send data over network via a remote I/O protocol
- E: time spent for computation
- P: total time spent for the application

The total time spent for the application via staging would be:

$$P_s = P_{in} + E + P_{out} \tag{1}$$

where

$$P_{in} = Rr_{in} + Ns_{in} + Wl_{in} + Rl_{in} \tag{2}$$

and

$$P_{out} = Wl_{out} + Rl_{out} + Ns_{out} + Wr_{out} \tag{3}$$

On the other hand, the total time spent for the application via remote I/O would be:

$$P_r = Rr_{in} + Nr_{in} + E + Nr_{out} + Wr_{out} \tag{4}$$

The actual computation time for both methods would be the same. The time difference in total (end-to-end) application time comes from network and disk I/O operations. For simplicity, we compare Input and Output operations separately.

In such a setting, for remote I/O to be more efficient than staging, we should have:

$$Rr_{in} + Nr_{in} < Rr_{in} + Ns_{in} + Wl_{in} + Rl_{in} \tag{5}$$

and

$$Nr_{out} + Wr_{out} < Wl_{out} + Rl_{out} + Ns_{out} + Wr_{out} \tag{6}$$

From Equation 5, we would get:

$$Nr_{in} - Ns_{in} < Wl_{in} + Rl_{in} \tag{7}$$

and similarly from Equation 6, we would get:

$$Nr_{out} - Ns_{out} < Wl_{out} + Rl_{out} \tag{8}$$

which means the time difference coming from using a specialized data transfer protocol versus a remote I/O protocol should be less than the overhead of extra read/write to the disk in staging. In other words, if your remote I/O library performs good in data transfer over network, or your local disk performance is slow, remote I/O might be advantageous over staging. Otherwise, staging method would perform better.

According to authors knowledge, Nrin and Nrout can not be measured separately. Since, these variables are plays important role in our initial mode, we would like to start with two separate models for each data access technique using our measurements with regression analysis in the following section.

3. A Realistic Model based on Regression

In order to define the model for the remote I/O vs staging data access techniques, a standard multiple regression analysis was performed using PASW Statistics 18 [1].

3.1. Regression Model for Data Staging

End-to-end performance of the application (Ps) is a dependent variable whereas Rrin, Nsin, Wlin, Rlin, Wlout, Rlout, Nsout, Wrout , DS (data size), AT(Access Techniques), NA (Network Architectures) were the independent variables for data staging technique model. Full, half, quarter, and eighth are grouped as the data size (DS). NA variable indicates the network architectures such as LAN, CAN, MAN, WAN1, WAN2, WAN3, WAN4, WAN5, WAN6, WAN7. AT variable which is data access techniques (sequential, jump, and random) indicates categorical type of variable. Therefore, (j-1) approach of regression, when including categorical variable in the model, applied [2]. It is also suggested by Kleinbaum [3] et al. when including a nominal independent variable since it can help to index categories of the nominal variable in regression analysis. This method is also called reference cell coding and it is about using dummy variables for one less of the number of categories (j-1). In this study, AT variable which has 3 categories is

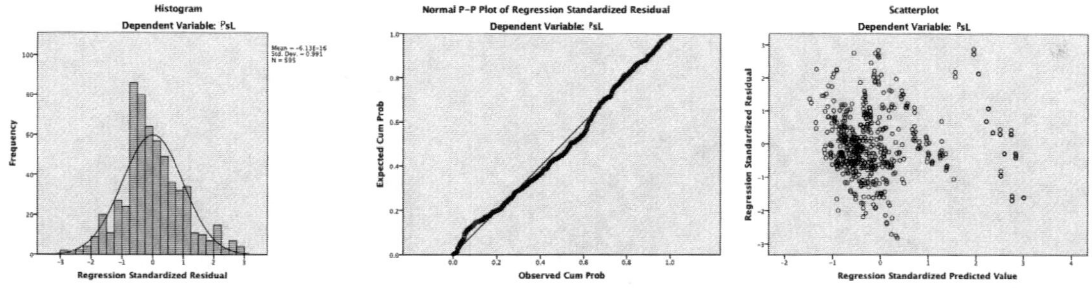

Figure 1. Staging Histogram After the Transformations

recoded into 2 dummy variables (ATj and ATr) that have values of 0 or 1. ATj dummy variable is for jump access technique and Atr dummy variable is for random access technique. The sequential access technique is not coded in a separate variable since it is defined when both ATj and ATr variables equal to 0. The intercept of the regression model will indicate the coefficient for the sequential category.

The initial investigation of the variables is showed that dependent variable Ps and other independent variables are not normally distributed.

Logarithmic transformation was used for variables Ps, Nsin, Wlin, Rlin, Wlout, Rlout, Nsout, and Wrout. Before logarithmic transformation, reflect method was applied for Rrin variable. Because of its relatively negative skewness reflect method helped to transform the distribution into the positive direction then logarithm applied. After the transformations all the skewness and kurtosis levels approached to the normal distribution values. After the logarithmic transformation skewness and kurtosis values improved. This was evident for the other variables. PASW Explore procedure was used to check each variables whether or not they improved in terms of normal distribution.

After transformation and removal of the outliers, residuals indicated normal distribution as it can be seen in Figure 1.

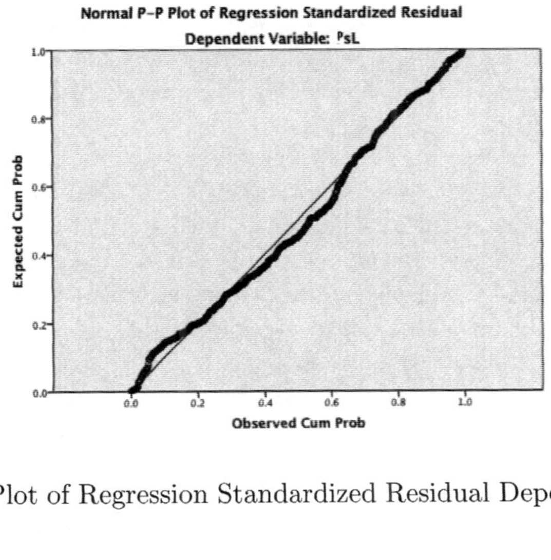

Figure 2. Normal P-P Plot of Regression Standardized Residual Dependent Variable PS After the Transformations

Figure 2 indicates that the residuals are normally distributed after the transformation.

Figure 3 shows the normality, linearity and homoscesdaticity assumptions. According to

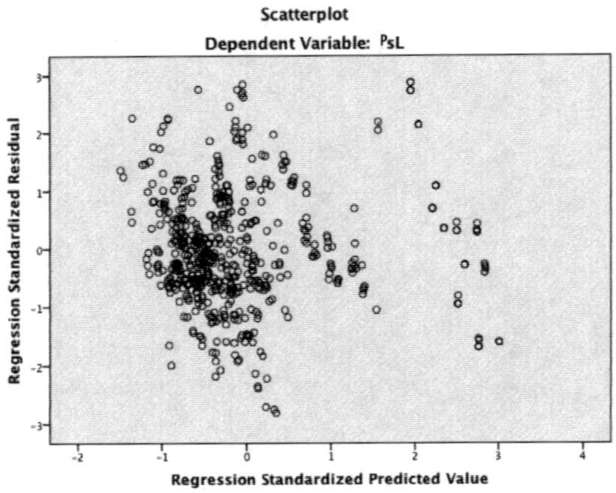

Figure 3. Staging Regression Standardized Predicted Values After the Transformations

Tabashnick and Fidell [4], if scatterplot of residuals with standardized predicted dependent variable values display the scores scattered around the zero line it indicates that the normality, linearity and homoscesdaticty assumptions are met. As values of the residuals scattered along the 0 line the assumptions were assumed to be met. Moreover, Tate [5] stated that moderate violations of the normality assumption may often be ignored with larger sample size as it does not affect the analysis negatively. Kleinbaum et al. also stated if normality assumption is not badly violated, the results reached by regression analysis will generally be reliable and accurate. In addition, Tabshnick and Fidell [4] noted that violations of linearity and homoscedasticity do not invalidate the analysis. Regarding the 595 observations, the sample was large enough and it was concluded that assumptions of regression analysis were met.

By using Mahalanobis distance with $p<.001$, multivariate outliers were checked for a number of variables in the model and no multivariate outliers are detected in the data. In fact, maximum Mahalanobis distance were 31.366 and it was less than the X^2 critical value. N= 595 and no missing data observed.

The results also indicated multicollinearity (high correlation among independent variables) was an issue. However, some of the independent variables were highly correlated each other: NsoutL and NsinL (.832) and , RlinL and RloutL (1.), WloutL was highly correlated with RlinL (.912) and with RloutL (.912).

When regression analysis was applied it was observed that Rlin variable is excluded from the analysis by the statistical program due to perfect correlation with the Rlout variable. Since Rlout and Rlin are identical in their relationship with the performance (dependent variable) no further solution was attempted to include Rlin in the model. Moreover, including perfectly correlated independent variables in the model causes greater standard errors and coefficients will be shown as not significant in the model.

The results indicate that an overall model of eleven predictors that significantly predict the performance in staging data technique access, R^2 =.984 and adjusted R^2= .983 $F(11,583)=3197.962$ $p<.001$. This model is accounted for 98 % of the variance in performance of staging data access technique. In other words, 98 % of the variance in the PsL was explained by Rrin, Wlin, Wlout, Nsout, Wrout , DS, AT, and NA.

Since there was a perfect relationship (1.) between the RlinL and RloutL, RlinL was excluded from the analysis by a statistical program due to the multicollinearity. Since the synthetic application creates the same amount of data, Rlout and Rlin operate the same amount of data. It is not surprising that both variables have strong relationship. T test statistics results indicate

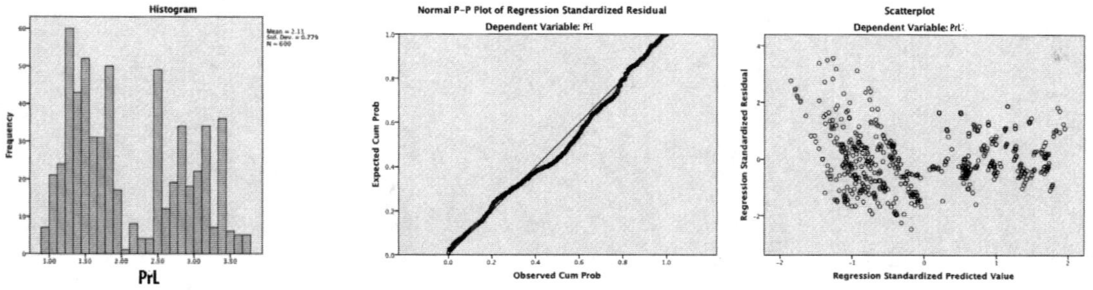

Figure 4. Remote I/O Histogram After the Transformations

that coefficients of the all independent variables are significant except WloutL. However, since multicollaniarity is normal with this type of data. It doesn't affect the inclusion of this coefficient in the regression equation.

PsL is estimated by the following regression model.

$$Ps = 1.133 + (-.240)(R_{rinL}) + .268(N_{sinL}) + .254(W_{linL}) + (-.009)(W_{loutL}) + (-.157)(R_{loutL}) \\ +(.269)(N_{soutL}) + .290(W_{routL}) + (.051)(D_S) + .003(N_A) + .054(A_{Tj}) + .154(A_{Tr})$$

3.2. Regression Model for Remote I/O

Performance of the application (Pr) was the dependent variable whereas Rrin, Nr, and Wrout, DS(data size), AT(Access Techniques), NA (Network Architectures) were the independent variables for remote data access techniques in distributed environments.

DS variable indicated transaction and data size including Full, half, quarter, and eighth (DS). NA variable includes the network architectures such as LAN, CAN, MAN, WAN1, WAN2, WAN3, WAN4, WAN5, WAN6, and WAN7. AT variable which were data access techniques (sequential, jump, and random) indicated categorical type of variable. Therefore (j-1) approach of regression or in other term reference cell coding, when including categorical variable in the model was applied in PASW.

The initial investigation of the variables showed that dependent variable Pr and other independent variables were not normally distributed. In fact, most of the variables were positively skewed. Logarithmic transformation was used for variables Pr, Rrin, Nr, and Wrout. After the transformations, all the skewness and kurtosis levels approached to the normal distribution values. Also, "L" is added to end of the variable name to indicate log(logarithmic) transformation. In perfectly normal distribution, skewness and kurtosis levels are supposed to be 0. Before the transformations, skewness and kurtosis for the dependent variable Pr were 2.845 and 7.046 respectively. After the log transformation, skewness and kurtosis values improved and became .360 and -1.270 respectively. PASW Explore procedure was used to check each variable whether or not they improved in terms of normal distribution.

Regression results indicated an overall model of nine predictors that significantly predict the performance in remote I/O data access technique. This model is accounted for nearly 98 % of the variance in performance of remote I/O data access technique. The coefficient of the intercept (constant) captures the coefficient of the sequential access technique as it was not included as a dummy variable in the equation because of the (j-1) dummy variable approach. The coefficient of jump access technique(Atj) is the difference in means between jump and sequential. Similarly, the coefficient of random access technique for the Atr is the difference in means between random and sequential.

$$Pr = .735 + (-.160)(R_{rinL}) + .527(N_{rL}) + .474(W_{routL}) + (.065)(D_S) + .011(N_A) + .080(A_{Tj}) + .322(A_{Tr})$$

4. Model Validation

Two real-life applications, Hurricane Data Archive and Blast, are used to test the validity of our model. In addition to these real-life applications, we have also used modified versions of our synthetic applications to test some extreme cases. The following sections provide the tests and their results.

4.1. Real-Life Applications

4.1.1. Data Archive for Coastal Science Coastal/Hurricane research group on CCT has been developing a Simulated Hurricane Database [6] hosted on Petashare[7] containing data produced from ADCIRC [8] application. Initially, archive database is populated by ADCIRC runs for hurricanes and tropical storms that have occurred in the Golf of Mexico over the past 50 years. Then, archive database provides information to additional application for hypothetical storm events. Some applications are available to analyze the simulated hurricane database We will use the application program to find out which remote data access technique is the best method.

The simulated hurricane database is around 38 GByte. Spider is used as an execution node, and eric is used as a data server node. So, we have used CAN network architecture for this experiment. We have staged in all the data files from eric to spider by globus-url-copy. After the execution, we have staged out all the files from spider to eric back in staging tests. We have repeated the tests five times to eliminate network affects. Application program has accessed the simulated hurricane database which is on eric with parrot/gsiftp remote I/O protocol. This application accesses small portion of the database, so we have used eighth data size. The application reads the database sequentialy.

Figure 5 provides both staging and remote I/O results. Remote I/O has better performance than staging. When we replace the variables with their corresponding values, the equation becomes:

$$Ps = 239.126 \tag{9}$$

When we replace the variables with their corresponding values, the equation becomes:

$$Pr = 92.64 \tag{10}$$

Our model also shows that remote I/O has better performance score. Real-life application confirms the validity of our model.

4.1.2. Blast Blast compares a sequence with those contained in nucleotide and protein databases by aligning the sequence with previously characterized genes. It finds regions of sequence similarity, which will yield functional and evolutionary clues about the structure and function of this sequence.

Blast is used for the second real-life application. Blast DNA database refseq_rna is used, it is around 1.3 GByte. Spider is used as an execution node and Queen Bee is used as a data server node (MAN). This application accesses all database, so we have used full data size The application reads the database sequentially.

According to the Figure 6, remote I/O also performing better then staging in our second real-life application.

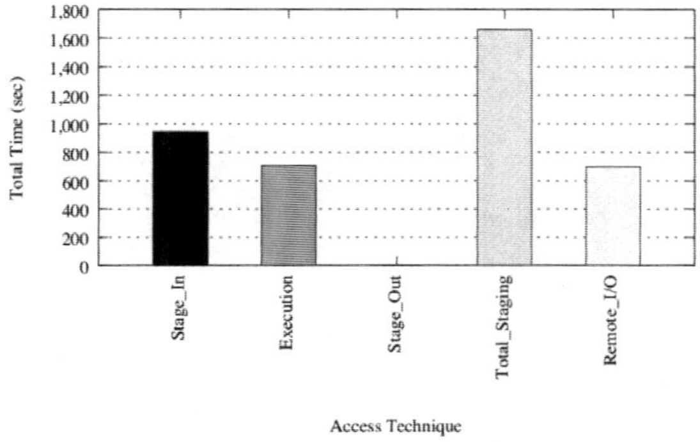

Figure 5. Simulated Hurricane Database Results

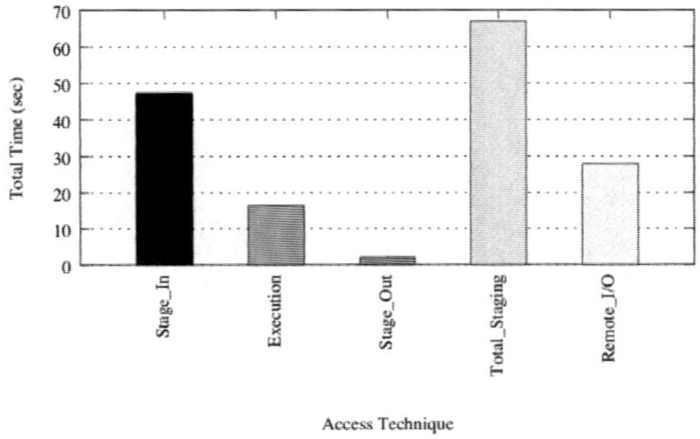

Figure 6. Blast Results

When we replace the variables with their corresponding values, the equation becomes:

$$Ps = 8.83 \tag{11}$$

When we replace the variables with their corresponding values, the equation becomes:

$$Pr = 5.84 \tag{12}$$

Our model also shows that remote I/O has better performance score. Real-life application confirms the validity of our model.

4.2. Synthetic Applications
We have used modified versions of our synthetic applications to test the validity of our models with different use cases. We will report two use cases in this section: full ratio and eighth ratio.

4.2.1. Synthetic Application with Full Ratio We have used our synthetic application and synthetic data. Spider is used as an execution node and Painter is used as a data server node (WAN4). This application accesses all data, so we have used full data size. The application reads the database randomly.

According to the Figure 7, staging is performing better then remote I/O in our first case synthetic application.

We applied our models as follows:

Staging

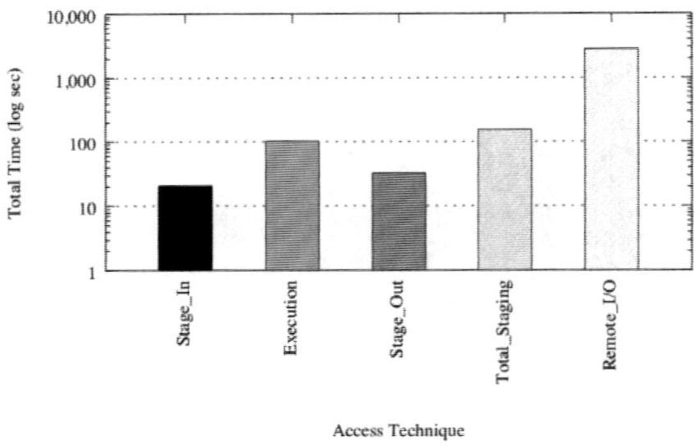

Figure 7. Synthetic Application Results with Full Ratio

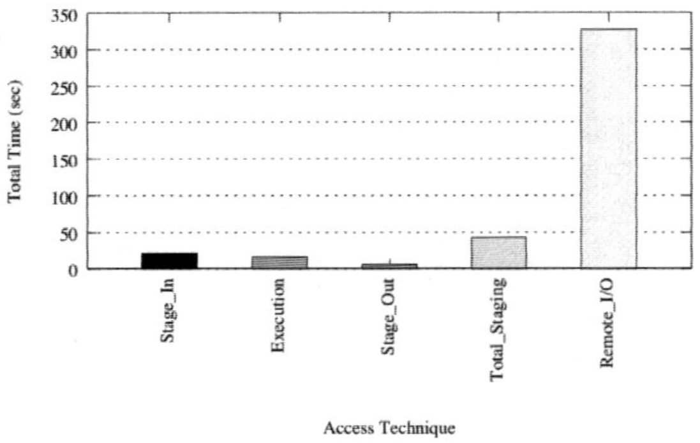

Figure 8. Synthetic Application Results with Eighth Ratio

When we replace the variables with their corresponding values, the equation becomes:

$$Ps = 12.70 \tag{13}$$

Remote I/O
When we replace the variables with their corresponding values, the equation becomes:

$$Pr = 1419.17 \tag{14}$$

Our model shows that staging has better performance score. Synthetic application confirms the validity of our model.

4.2.2. Synthetic Application with Eighth Ratio We have used our synthetic application and synthetic data. Spider is used as an execution node and Painter is used as a data server node (WAN4). This application accesses eighth of the data, so we have used eighth data size. The application reads the database randomly.

According to the Figure 8, staging is performing better then remote I/O in our second case synthetic application.

We applied our models as follows:

Staging When we replace the variables with their corresponding values, the equation becomes:

$$Ps = 5.61 \tag{15}$$

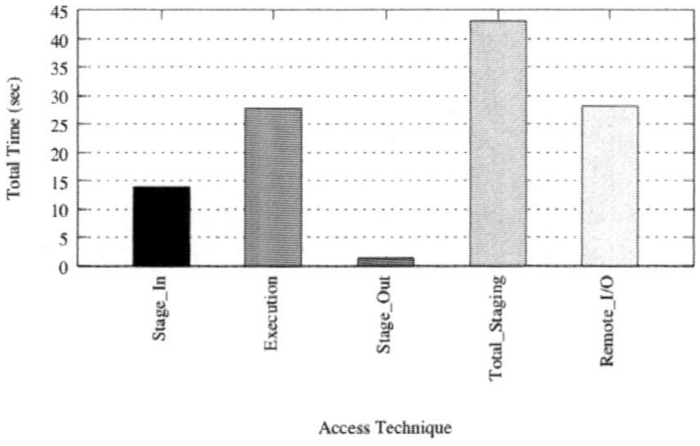

Figure 9. Extreme Case Synthetic Application Results with Full Input 1/100 Output Sequential

Remote I/O When we replace the variables with their corresponding values, the equation becomes:

$$Pr = 164.585 \tag{16}$$

Our model also shows that staging has better performance score. Synthetic application confirms the validity of our model.

4.3. Extreme Cases

We have used modified versions of our synthetic application to test some extreme use cases as well. We will report two extreme use cases in this section. For each case, we run the simulation for two different data access techniques (sequential and random). On the first case, our simulation reads all data and produces 1/100 output data ratio. On the second case, our simulation reads 1/100 input data ratio and produces the same amount of data.

4.3.1. Full Ratio Input, 1/100 Ratio Output We have used our synthetic application and synthetic data. Spider is used as an execution node and Quinbee is used as a data server node (MAN). This application accesses all data, so we have used full data size.

According to the Figure 9, remote I/O is performing better then staging in our first extreme case synthetic application.

We applied our models as follows:

Staging When we replace the variables with their corresponding values, the equation becomes:

$$Ps = 6.50 \tag{17}$$

Remote I/O

When we replace the variables with their corresponding values, the equation becomes:

$$Pr = 5.45 \tag{18}$$

Our model shows that remote I/O has better performance score. Synthetic application confirms the validity of our model.

According to the Figure 10, staging is performing better then remote I/O in our first extreme case synthetic application with random data access.

We applied our models as follows:

Staging When we replace the variables with their corresponding values, the equation becomes:

$$Ps = 5.78 \tag{19}$$

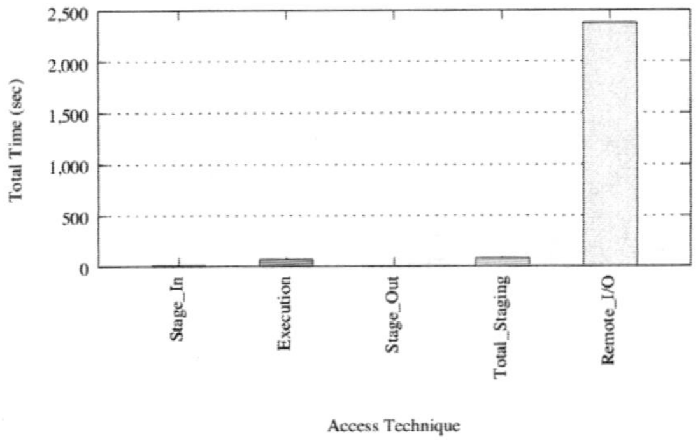

Figure 10. Extreme Case Synthetic Application Results with Full Input 1/100 Output Random

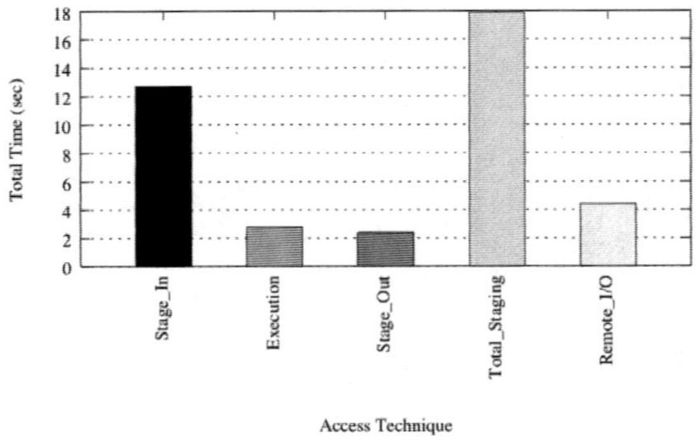

Figure 11. Extreme Case Synthetic Application Results with 1/100 Input 1/100 Output Sequential

Remote I/O When we replace the variables with their corresponding values, the equation becomes:

$$Pr = 1296.52 \tag{20}$$

Our model shows that staging has better performance score. Synthetic application confirms the validity of our model.

4.3.2. 1/100 Ratio Input, 1/100 Ratio Output We have used our synthetic application and synthetic data. Spider is used as an execution node and Quinbee is used as a data server node (MAN). This application accesses 1/100 data, so we have used eighth data size.

According to the Figure 11, remote I/O is performing better then staging.

We applied our models as follows:

Staging

When we replace the variables with their corresponding values, the equation becomes:

$$Ps = 3.50 \tag{21}$$

Remote I/O

When we replace the variables with their corresponding values, the equation becomes:

$$Pr = 2.51 \tag{22}$$

Our model shows that remote I/O has better performance score. Synthetic application confirms the validity of our model.

According to the Figure 10, staging is performing better then remote I/O with random data access.

We applied our models as follows:

Staging

When we replace the variables with their corresponding values, the equation becomes:

$$Ps = 4.13 \tag{23}$$

Remote I/O

When we replace the variables with their corresponding values, the equation becomes:

$$Pr = 162.50 \tag{24}$$

Our model shows that staging has better performance score. Synthetic application confirms the validity of our model.

5. Related Work

According to Stockinger [9], the entire resource selection problem requires detailed cost models with respect to data transfer. A cost model for data-intensive applications is discussed in [10] where theoretical models for data intensive job scheduling are presented. The metric for measuring efficiency is the effective time seen by the client application. The model includes all important factors in a distributed Data Grid and takes various storage and access latency into account to determine optimal data access.

More general performance engineering approaches are discussed in [11]. In this work, they analyze a typical distributed system and point out performance analysis aspects in order to improve the overall job execution time of the system. They give a detailed look at the following two domains: data access and networking.

Data staging and remote I/O have been compared by different studies from different aspects. For instance, GridFTP as a staging technique and RFIO as a remote I/O technique have been compared by Kalmady and Tierney [12] on wide area networks. According to Kalmady and Tierney, the performance of RFIO is better than gsiftp for one stream. Gsiftp performs better on multiple streams. With tuning, RFIO becomes pretty close to gsiftp which means that proper tuning can make the difference. They came up with the following observations: i) setting the TCP buffer size a proper value is the most important factor for a good performance; ii) 2-3 parallel streams will gain an additional 25% performance over a single tuned stream; iii) a mechanism of dynamically varying the buffer sizes during data transfer is needed, because of sensitivity of the variation in network traffic; iv) the same throughput can be gained with tuned buffers using untuned TCP buffers with enough parallel streams.

Thain and Livny compared performance of parrot/chirp with other staging and remote I/O techniques on the study [13] with Andrew-like benchmark. According to the authors, separating computation from storage makes I/O cost high. Copying data gets slower over the network, but the slowdown in the network makes staging acceptable because of increased throughput via remote parallelization. Also, authors point out that the differences in the performance between Chirp, ftp, and NeST are because of the cost of the metadata lookups.

Modeling the data access techniques in a distributed environment is not a well-studied area. Staging the data before it is needed can be very complicated because of the dynamic nature of the distributed environment such as real-time network traffic and congestion. Limitations of bandwidth, storage space at certain sites, and data specifications can affect the staging of the data. A mathematical model for a basic data staging problem is studied by Theys et al.

[14]. All parameter values for the network and data request a stay fix in the scheduling process. However, the parameter values for the model are changed temporarily to reflect the dynamic nature of the distributed environment. The parameters can be categorized by the node storage capacity and node number, the link availability starting and ending time, bandwidth, latency, source node, and destination node. Also, every request has data size, list of sources, and list of destinations. Elwasif et al. [15] developed a model for farming applications with and without server side staging to analyze the effect of staging, and verify the model with experiments and simulations.

A remote I/O performance model [16] can estimate remote I/O cost before performing the application, so the application can be evaluated better. The paper also presented the design of a remote I/O performance predictor that gives the user a concept how much remote I/O costs for the application, so the appropriate parameters can be chosen for the application. A practical remote data access model is presented in [17] which describes a tightly coupled, a data oriented infrastructure approach with building a leading edge technology to provide very high-speed, and widespread access to large data storage. Antoniu et al. [18] provide a transparent data access model which enables users to access data via global identifiers. This model manages data persistence dynamically and transparently in distributed environment using two approaches which are distributed shared memory and peer to peer systems.

Shivam et al. [19] present the Non-Invasive Modeling for Optimization (NIMO) system which automatically learns cost models for predicting the execution time of computational science applications on distributed environments. NIMO first generates training samples for distributed applications, then by using these samples learns cost models with statistical learning techniques. NIMO is an active system, so it deploys and monitors the sampled application under different conditions. NIMO is also non-invasive so it collects training data from passive streams without effecting not only the operating system or the application, but also the application source, or library.

On the other hand, there are some challenges that arise based on NIMO. According to Shivam et al. [19], sampling acquisition may have high overhead. Also, the number of samples needed, given the level of accuracy, increases exponentially. The training sample set may not represent the entire operating range of the system.

Although our approach and NIMO can suggest which data access method to use, there are significant differences between them. Researchers can use our approach before the design phase, so they can design the application based on the best data access technique. NIMO does not only evaluate the data access methods, but also the active system elements such as CPU, cache, and I/O system behaviors. Sometimes, NIMO can prefer a data access technique which is worse than others in terms of overall throughput, but preferable in terms of other active elements such as CPU, cache, and I/O.

6. Conclusions and Future Work

We have developed generic models and set guidelines for the application developers which would help them to choose the most appropriate data access method for their application. We defined the parameters that potentially affect the end-to-end performance of the distributed applications which need to access remote data.

Sequential and jump data access always performs better than random data access on all network architectures. High-speed networks improve the data transfer ratio dramatically. Random data access on remote I/O is not benefiting from high-speed network architectures, so the performance decreases dramatically. The differences on sequential and jump is less than the differences on random. So, researcher should be more careful when remote I/O was chosen. Since all data should be staged in before the execution, decreasing the data ratio improves the remote I/O performance.

Journal of Physics: Conference Series **540** (2014) 012003 doi:10.1088/1742-6596/540/1/012003

Increasing the distance between execution node and data server node with high-speed network decrease the the gap between sequential and jump performance. Without high-speed network, it decreases the gap between remote I/O and staging on all data access techniques. On the other hand, it increases the gap on smaller data ratios. Also, choosing proper remote I/O protocol is a crucial decision for the end-to-end application performance.

References

[1] 2010 www.spss.com URL `www.spss.com`
[2] Norusis M J 2008 *SPSS 16.0 Guide to Data Analysis* (prentice hall Inc.)
[3] Kleinbaum D G 1998 *Applied Regression Analysis and Other Multivariable Methods* (Duxbury Press)
[4] G T B and S F L 2007 *Using multivariate statistics* (Pearson Education Inc.)
[5] Tite R 1996 General linear model applications
[6] Bhagawaty R S H, Jiang L, Dubbaka S, Hu K, Pothanis S, Allen G, Brener N, Chen Q and Kosar T 2010 *CLADE*
[7] 2010 http://www.petashare.org/ URL `http://www.petashare.org/`
[8] http://adcirc.org/ URL `http://adcirc.org/`
[9] Stockinger H May 2005 Data management in data grids - habilitation overview Tech. rep. Research Lab for Computational Technologies and Applications
[10] Stockinger H, Stockinger K, Schikuta E and Willers I February, 2001 *9th Euromicro Workshop on Parallel and Distributed Processing (PDP 2001), IEEE Computer Society Press, Mantova, Italy*
[11] Laure E, Stockinger H and Stockinger K 2005 *Journal of Concurrency and Computation: Practice and Experience, Wiley Press* **17(2-4)**
[12] Kalmady R and Tierney B 2001 *Proceedings of Computers in High Energy Physics*
[13] Thain D and Livny M 2003 *Univ. of Notre Dame, Computer Science and Engineering Dept.*
[14] Theys M D, Siegel H J, Beck N B, Ta M and Jurczyk M 1998 *Proceedings of the Seventh Heterogeneous Computing Workshop*
[15] Elwasif W, Plank J and Wolski R 2001 *Proceedings of the 1st International Symposium on Cluster Computing and the Grid*
[16] Shen X, keng Liao W and Choudhary A 2001 *IASTED Applied Informatics, Innsbruck, Austria*
[17] Andrews P M, Sherwin T and Banister B 2003 *Proceedings on 20th IEEE/11th NASA Goddard Conference*
[18] Antoniu L G and BougB M J 2005 *Scalable Computing: Practice and Ezperience*
[19] Babu S, P Shivam and Chase J 2006 *International Conferance on Very Large Data Bases (VLDB)*

Implementation of a solution Cloud Computing with MapReduce model

Chalabi Baya
Higher National School of Computer Science ,Algeria

E-mail: b_chalabi@esi.dz

Abstract. In recent years, large scale computer systems have emerged to meet the demands of high storage, supercomputing, and applications using very large data sets. The emergence of Cloud Computing offers the potentiel for analysis and processing of large data sets.
Mapreduce is the most popular programming model which is used to support the development of such applications. It was initially designed by Google for building large datacenters on a large scale, to provide Web search services with rapid response and high availability.
In this paper we will test the clustering algorithm K-means Clustering in a Cloud Computing. This algorithm is implemented on MapReduce. It has been chosen for its characteristics that are representative of many iterative data analysis algorithms. Then, we modify the framework CloudSim to simulate the MapReduce execution of K-means Clustering on different Cloud Computing, depending on their size and characteristics of target platforms.
The experiment show that the implementation of K-means Clustering gives good results especially for large data set and the Cloud infrastructure has an influence on these results.

1. Introduction

In recent years, we are witnessing the development of Applications that deal with very large data sets (about Peta bytes) and distributed throughout the world. [9]. These applications require infrastructure that offer particularly wide storage capacity and processing capabilities very important[9]. In addition to these fundamental characteristics, the communication aspect is also important insofar as these applications require of high speed networks and secure. To respond to Such requirements, which are out of reach for people individual, a new class of architectures has seen the day. This category includes grid architectures [14] and Cloud architectures [3]. The expansion of these architectures is done both in the field of industry and in the research science. On the industrial side, companies such as Google [1] have introduced large data centers on a global scale to provide web services that are both stable and highly available, ensuring fast response times for users. In the academic field, many scientific research projects have been launched to provide solutions to manage large data sets in a very broad scale [4]. Research conducted within the framework of these differents projects have led to the emergence a new concept in computing, namely the Cloud Computing [3]. The interest in this new concept led many companies specializing in computer to devote substantial budgets for make it available to users. Among the products that have been placed on the market, we can cite Google App Engine [1] provides an overview of the MapReduce technique and a brief introduction to Hadoop. In Section 4, we presentCloudSim simulator. Section 5 is allocated for description of the data mining algorithm that requires important data set which is k-means clustering. While Section 6 presents the implementation of K-means Clustering in the Cloud Computing by using Hadoop.

2. Related work

Currently k-means Clustering can be implemented in cloud frameworks for iterative algorithms . Zhao et al [23] adapt this algorithm for Hadoop MapReduce. The work focuses on implementing the K-mans clustering with the read-only convergence heuristic in the Hadoop MapReduce pattern . However, as Hadoop does not support iterative processing by design, this implementation has little choice regarding data affinity even with the read- only convergence heuristic . Twister [15] and HaLoop [25] add modifications to MapReduce for supporting iterative computation. In ower work we implement k-means with Hadoop, but with different infrastractures that used by Zhao et al[23]; also in the experiments we give the memory size used and network consumption by the benchmarks which are not done in others works. Based on these experiments, we have changed the framework CloudSim for supporting the simulation of MapReduce execution to see the effect of the availability of virtual machine (VM)on the execution time.

3. MapReduce model

In this section, we present a brief introduction of MapReduce. MapReduce is a partitioning mechanism of tasks for distributed execution on a large number of servers [16]. It is primarily intended to tasks such as batch. Its principle is summarized as follows: it is to break a task into smaller tasks, or more precisely cut a task involving very large volumes of data on identical tasks with subsets of the data. Tasks (and their data) are then dispatched to different servers, then results are retrieved and consolidated. The upstream phase, decomposition of tasks is called Map part, while the downstream phase, the consolidation of the results is called the party Reduce. The principle is relatively simple but the contribution of MapReduce is to well conceptualize view to normalizing the operations of stewardship so that same framework can support a variety of partitionable tasks, allowing developers to focus on the treatments themselves, while The framework supports the logistics of distribution.
Operations "Map" and "Reduce" are general and have a simple interface. Each one receives a sequence records (records within the meaning of the model), and each usually produces output record. A record consists of a key and a value. The input records presented to the mapper by its appellant, does not guarantee or ordering relation for the execution of Map tasks. The work of mapper is to create (or not) a number of records in response to each input record. Records are presented to reducer by its caller and are placed in a cell according to their key, so that all records with a same key are presented as a package to the reducer. The reducer then examines packets sequentially using an iterator.

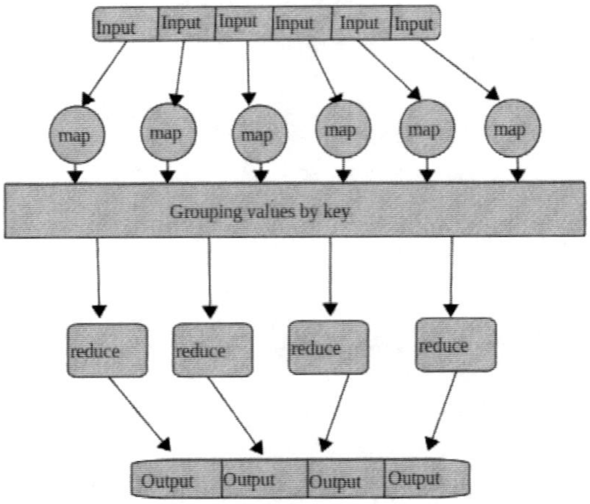

Figure 1. Operating principle of MapReduce

3.1. Hadoop

The implementation of MapReduce by Apache Hadoop is coupled with the distributed file system (HDFS). It is similar to the architecture of Google MapReduce runtime in which accesses data to HDFS [2], all mape local disks of the compute nodes in a file system single hierarchy, thus allowing the replication of calculation data between all nodes. HDFS also replicates data on multiple nodes so that all failures nodes containing a portion of data does not affect the calculations that use these data. Hadoop order the MapReduce computation tasks according to data locality, thereby improving the overall strip width. the outputs mapped tasks are first stored in local disks to perform later access to operation reduce[24] over HTTP connections. Although this approach simplifies the mechanism tasks processing, it generates a significant communication overhead for intermediate data transfers, especially for applications that frequently produce small intermediate results. The main mean used by hadoop to achieve fault tolerance is to complete execution (or rerun) the assigned tasks in the cluster [5]. Another special feature that it is useful in Hadoop mentioning is the speculative execution model. In a parallel system, there is the problem of the presence of slower nodes that slow the calculation of the entire cluster. In Hadoop, since the tasks are executed isolated from each other, then the same input may be treated several times in parallel by different nodes.

4. CloudSim

CloudSim is a framework that provides a simulation generalized and extensible allowing modeling, simulation and experimentation of new infrastructure of Cloud Computing and application services involved. It covers most of the activities that take place within a Data Center in detail. It allows [21]:

- Simulate the definition of physical data centers in terms of physical machines consisting of processors, storage devices, memory and internal bandwidth.
- Simulate the virtual machine specification, their creation and destruction. Management of virtual machine, allocation physical hardware resource for the operation of virtual machines based on differents policies.

Journal of Physics: Conference Series **540** (2014) 012004 doi:10.1088/1742-6596/540/1/012004

5. Proposed implementation

The K-means method [11] is known for its degradation when all the data growing in terms of number of objects and dimensions [17, 18]. To address this requirement, we implement this algorithm with MapReduce. In this section, we describe the clustering algorithm k-means clustering with MapReduce model.

(i) The set of input data (file of data points) is partitioned into N portions. Each part is sent to a Maper.

(ii) In the Map function, the distance (Euclidean distance) between each point and each cluster center is calculated and each point is labeled with the index of the center cluster where the distance is the smallest. The outputs of the function Map is the key-value pair:
   - Key: cluster ID.
   - Value: the coordinates of the point (recording).

(iii) The function "Map" produced a large amount of data. So, we use a function Combine to reduce the size before sending it to the function Reduce. Combine function calculates the average coordinates for each cluster ID, and the number of records. It outputs a pair key-value for each cluster:
   - The Key: Custer ID.
   - Value: the number of records and the values average coordinates.

(iv) All data points in the same cluster are sent to one Reducer. In the Reduce function, new coordinates of the centers of clusters are easily calculated (same treatment as the Combine function except that the Combine function is executed at the Mapers nodes while Reduce is executed in Reducers nodes). The output of the reducer is:
   - Key: cluster ID.
   - Value: new coordinates of the center of cluster.

(v) The new coordinates of the centers of clusters are compared to the original. To do this, we need to save the centers of clusters of the previous iteration. If the difference does not exceed a threshold predefined, then the program ends, and we found clusters. Otherwise, set the centers of clusters newly generated and we repeat steps 2 to 5.

6. Experements with Hadoop and results

To implement K-means clustering with MapReduce we use Hadoop and programming language Java is chosen, we experience this algorithm in a Cloud Clustering which constitutes five machines where the hardware specifications of the master node (jobtracker) are:

- Memory: 1GB.
- processor :2.8 GHz.

For the four virtual machines, we have used virtualbox and the client nodes (tasktracker) hardware specifications are:

- Memory: 512 MB.
- processor : 1.4 GHz.

The OS is Ubuntu-10.10.
The benchmarks used to make the experiments of K-means Clustering with Hadoop are generated aleatory and the number of clusters K is equal to 9. Also 9 centers of initial clusters are randomly generated.
The Size of benchmarcks in MB are respectively :5.70, 28.50, 56.90, 170.70, 284.50. Ganglia [12] is used for extracting the metrics. We will retrieve metrics values for an iteration execution of K-means clustering. The following figures show the results of experiments (execution time, memory size consumed, consumption network):

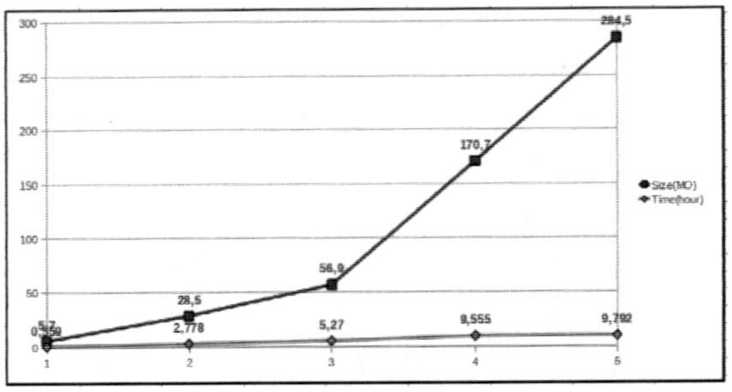

Figure 2. Execution time of five benchmarks used for K-means clustering in the Cloud Clustering .

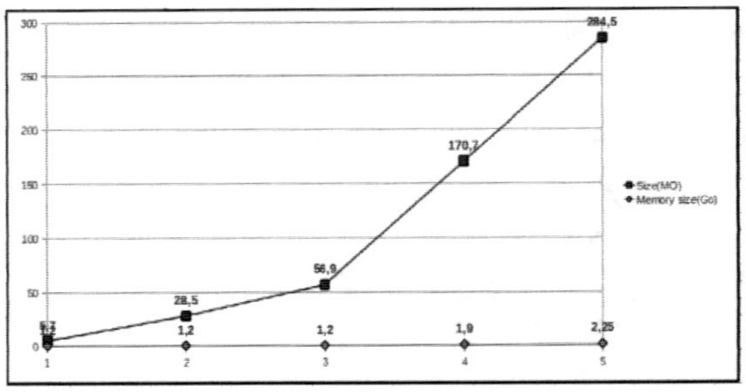

Figure 3. Memory size used by the five benchmarks used to K-means Clustering in the Cloud Clustering.

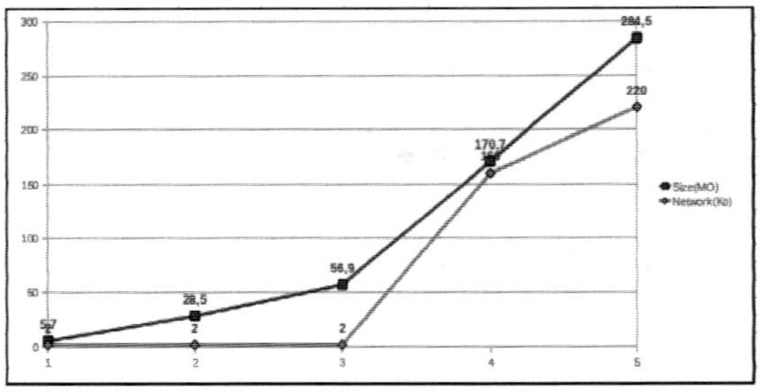

Figure 4. Network consumption by the five benchmarks used to K-means Clustering in the Cloud Clustering.

6.1. Discussion

Based on these results, we note that for the three first experiments, the execution time increases almost with the same value as the size of the data handled because these three experiments are executed at the same node which is the master node (one task) as the block size used by HDFS file is 64MB and data used are all (5.7,28.5, 56.9) less than 64MB. For the last two experiments, the execution time is almost the same despite that the size of the data manipulated in the last experiment and almost double that of the fourth experiment, because to the fourth iteration

the number of tasks Map is 3 and the three tasks are executed in parallel then the task Reduce (we choose to use a single Reducer in all experiments). For the last experiment the number of tasks Map is 5 and since we have a sufficient number of nodes, all tasks are processed in parallel. Parallel execution shows that we just near about the same execution time for recent experiments (the time difference is due to data locality) for the fourth experimentation, all tasks using local data (3 replies) but for the last experiment it is not all tasks that access their data locally.
The second finding is that, compared to the execution time of the third experiment with the execution time of the fourth and fifth experiment, we notice that there is a big difference (although the size of data processed in the first experiment is 56.9 MB and the data size of each task Map addressed in the fourth and fifth experiment is 64 MB. These tasks are performed in parallel. And since the task of the third experiment is performed on the master node and the tasks of the fourth and fifth experiment are treated at the client nodes which have different physical characteristics of the the master node. This implies that the material characteristics have an influence on the execution time, is to say more than the memory size is large and the speed processor is fast, the execution time is good.

7. Experimenting with cloudsim and results

To see the effect of the physical characteristics on the execution time, we simulate the performance of K-means clustering on different Clouds characteristics. For Data-center model used (IaaS), we used Hosts with the following characteristics:

- Ram :3000 Mo.
- Bw = 1000.
- Storage = 1000000 MO.

For processors, they are modeled in terms of MIPS (millions of instructions per second). We used a Pes with a value of Mips égalle 1000. In all experiments the number of virtual machines is 20, we have used virtual machines VirtualBox to take the same type as that used in the experiments with Hadoop. We used different models:

Table 1. Size memory and processor MIPS used for each expérimantation.

	Memory Size (MO)	Processor Mips
1	512	140
2	1024	280
3	1536	420
4	2048	560

The data sizes processed in GB are: 0.312, 0.937, 1.937, 2.812.
For each data size, we will allocate experiments with four different characteristics of virtual machine (RAM, CPU).

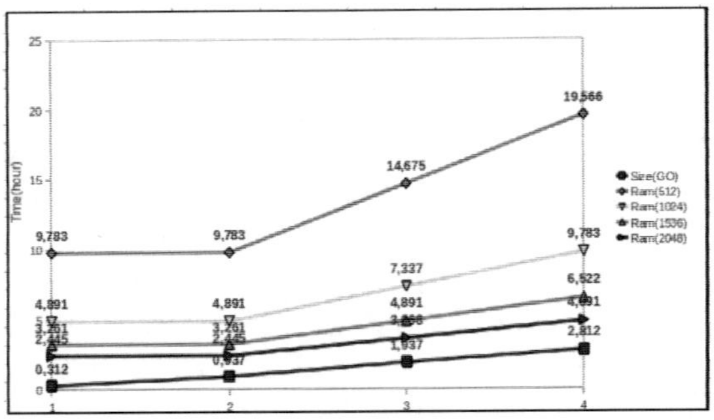

Figure 5. Variation of the execution time under different configurations (memory, CPU).

7.1. Discussion

Figure 5 shows the execution time is inversely proportional with the capacity of the RAM used by the nodes and processing capacity. Calculation of the gain between the four experiments, ascending Mips and Ram is as follows:

Table 2. varying the gain of execution time as a function of the size of the ram and the speed of processor.

Taille(GO)	Gain of time between different experiments		
	Between(1 & 2)	Between(2& 3)	Between(3& 4)
0.31	2.05	1.499	1.480
0.94	2.05	1.499	1.480
1.94	2.00	1.500	1.333
2.81	2.00	1.500	1.333

For the first and the second experiment, the size of data is triple that used in the first. By cons, we note that the execution time remains the same, that the size of each task Map is 64Mo and the number of tasks for each size is 5 and 15, respectively, and each task is modeled by one cloudlet. The cloudlets are executed in parallel before pass their results to cloudlets Reducer that beginning their treatment after the cloudlets that model Map tasks are completed.
For the last two, the execution time is different since the number of virtual machines needed to perform their tasks (Map) is not available (the number Task (cloudlets) is respectively 30 and 45) for the fourth and fifth experiment, and the number of available virtual machines is 20, and since each task (cloudlet) must be executed by a VM. This implies that Not all tasks are run in parallel.
The table 2 shows that gain degree is different, it does not increase by the same amount even though the added power is the same.

8. Conclusion

The popularity of K-means algorithm is due to its procedure simple and rapid convergence to a suitable solution but K-means is known for its degradation when the data set grows. MapReduce model is more appropriate for this type of algorithms that require large power storage and computation. We are implementing K-means with MapReduce (Hadoop) which is fault tolerant and given the opportunity to run speculative. The results of our experiments showed that cloud clustering is most effective when the data set is very large. Plus size data large, implying a fairly

Journal of Physics: Conference Series **540** (2014) 012004 doi:10.1088/1742-6596/540/1/012004

large number of tasks Map, the more consumption of resources (memory, network) increases. Based on simulations with CloudSim we can conclude that the number of available nodes and power storage and computing have an influence on the execution time. If the number of nodes required for the tasks Map is available, then the gain of the execution time is proportional to the power of nodes used and this gain degree is different, it does not increase by the same amount even though the added power is the same.

References

[1] Google app engine. `http://appengine.google.com`, Juillet 2008.

[2] Apache. hdfs architecture. `http://hadoop.apache.org/common/docs/current/hdfs_design.html`, 2009.

[3] Amazon elastic compute cloud (amazon ec2). `http://aws.amazon.com/ec2/.`, 2010.

[4] Spring.net. `http://www.springframework.net.`, 2010.

[5] A. Rasin D. J. Abadi-D.J. DeWitt S. Madden An. Pavlo, E. Paulson and M. Stonebraker. A comparison of approaches to large-scale data analysis. In *Proceedings of the 35th SIGMOD international conference on Management of data*, pages 165–178, 2009.

[6] B. Krishnamurthy C. Graham. Key differences between web 1.0 and web 2.0. *First Monday 13.6*, 2008.`http://www.uic.edu/htbin/cgiwrap/bin/ojs/index.php/fm/article/view/2125/1972.`

[7] Y. An Lin-Y.Yuan Yu G. Bradski A. Ng K.Olukotun C-T. Chu, S. Kyun Kim. Map-reduce for machine learning on multicore. *NIPS*, pages 281–288, 2006.

[8] M. Mattess R. Buyya C. Vecchiola, X. Chu. *Aneka - Integration of Private and Public Clouds.* -13: 978-0470887998. Wiley Press, New York, USA, 2010.

[9] Raphael Y. De Camargo, Andrei Goldchleger, and Fabio Kon. *InteGrade: a tool for executing parallel applications on a Grid for opportunistic computing.* Fortaleza-CE, 2005.

[10] C. Grzegorczyk et al D. Nurmi, R. Wolski. Eucalyptus: A technical report on an elastic utility computing architecture linking your programs to useful systems. Technical report, University of California Santa Barbara Computer Science, Santa Barbara, 2008.

[11] V. Faber. Clustering and the continuous k-means algorithm. In *Los Alamos Science*, volume vol. 22, pages pp. 138–144, 1994.

[12] M.L. Massie F.D. Sacerdoti, M.J. Katz and D.E. Culler. Wide area cluster monitoring with ganglia. In Citeseer, editor, *In Proceedings of the IEEE Cluster 2003 Conference*, 2003.

[13] Y. Liu R. Buyya H. Sun, J. Huai. Rct: A distributed tree for supporting efficient range and multi-attribute queries in grid computing. In *Elsevier Science*, volume 24 of *0167-739X*, page 631–643. Amsterdam, Juillet 2008.

[14] C. Kesselman I. Foster. The grid: Blueprint for a new computing infrastructure. *Morgan Kaufmann*, page 677, 1998.

[15] B. Zhang T. Gunarathne S.-H. Bae J. Qiu J. Ekanayake, H. Li and G. Fox. "twister: a runtime for iterative mapreduce,". In Eds. ACM, editor, *in HPDC, S. Hariri and K. Keahey*, page pp. 810–818, 2010. [Online]. Available: http://dblp.uni-trier.

[16] C. Dyer J. Lin. Data-intensive text processing with mapreduce. *Morgan Claypool*, pages 102–105, Avril 2010.

[17] M. Joshi. Parallel k-means algorithm on distributed memory multiprocessors. Technical report, University of Minnesota, 2003.

[18] P. Sattayatham K. Kerdprasop, N. Kerdprasop. Weighted k-means for density-biased clustering, lecture notes in computer science. In *Lecture Notes in Computer Science*, pages 488–497. Data Warehousing and Knowledge Discovery (DaWaK), 2005.

[19] M. Kunze L. Wang and J. Tao. Performance evaluation of virtual machine based grid workflow system. *Concurrency and Computation: Practice and Experience*, 20(15):1759–1771, 2008.

[20] A.Beloglazov C. A. F. Rose-R. Buyya N. Rodrigo. Calheiros, R.Ranjan. Cloudsim: A toolkit for modeling and simulation of cloud computing environments and evaluation of resource provisioning algorithms. *Wiley Press*, 41(1):23–50, January 2011.

[21] D. Karunamoorthy R. Buyya R. N. Calheiros, C. Vecchiola. Aneka platform and qos-driven resource provisioning for elastic applications on hybrid clouds. *Future Generation Computer Systems*, 2011.

[22] M. Kunze V. B ge, Y. Kemp and G. Quast. Application of virtualisation techniques at a university grid center. *e-Science*, 2006.

[23] H. Ma W. Zhao and Q. He. "parallel k-means clustering based on mapreduce,". In *Cloud Computing, ser. Lecture Notes in Computer ScienceM. Jaatun, G. Zhao, and C. RongEds. Springer Berlin / Heidelberg, 2009.*

[24] T. White. *hadoop:the definitive guide.* 978-1-4493-8974-1. O'Reilly Book Evangelist, 2011.

[25] M. Balazinska Y. Bu, B. Howe and M. D. Ernst. "haloop: Efficient iterative data processing on large clusters,". In *36th International Conference on Very Large Data Bases*. Singapore, September 14–16, 2010.

Journal of Physics: Conference Series **540** (2014) 012005 doi:10.1088/1742-6596/540/1/012005

Green Cloud Computing: An Experimental Validation

Rogerio Castellar Monteiro, M. A. R. Dantas
Research Laboratory of Distributed Systems (LaPeSD)
Department of Informatics and Statistic (INE)
Federal University of Federal Santa Catarina (UFSC)
88040-900 – Florianópolis, SC – Brazil

E-mail: rogerio@inf.ufsc.br, mario.dantas@ufsc.br

Martius Vicente Rodriguez y Rodriguez
Programa de Pós-graduação em Administração - PPGAd
Faculdade de Administração e Contabilidade - EST
Universidade Federal Fluminense - UFF
Rua São Paulo, 30, Centro
24020-005 - Niteroi, RJ - Brazil

E-mail: matius@kmpress.com.br

Abstract. Cloud configurations can be computational environment with interesting cost efficiency for several organizations sizes. However, the indiscriminate action of buying servers and network devices may not represent a correspondent performance number. In the academic and commercial literature, some researches highlight that these environments are idle for long periods. Therefore, energy management is an essential approach in any organization, because energy bills can causes remarkable negative impacts to these organizations in term of costs. In this paper, we present a research work that is characterized by an analysis of energy consumption in a private cloud computing environment, considering both computational resources and network devices. This study was motivated by a real case of a large organization. Therefore, the first part of the study we considered empirical experiments. In a second moment we used the GreenCloud simulator which was utilized to foresee some different configurations. The research reached a successful and differentiated goal in presenting key issues from computational resources and network, related to the energy consumption for real private cloud.

1. Introduction

Cloud configurations are steadily increasing their energy consumption due to the large number of clusters and multi-clusters environments. On the other hand, it is estimated that the waste of energy to support these cloud configurations are much higher than costs to acquire these configurations.

An energy consumption policy is an essential approach to be considered in any organization. The lack of knowledge on how to adopt standards, and also metrics, leads many data centers managers to a puzzle on how to measure energy efficiency from their environments. In other words, in many cases, managers do not know how to calculate their configurations energy efficiency.

The importance of this research is exactly in this key issue for any organization that has, or utilizes, a cloud configuration. The contribution reflected a real case study from a large corporation. This study case represents a private cloud scenario, where not only computational resource facilities are taken in consideration, but, also, costs related to the transmission from network devices. Therefore, the differential contribution of this article is to provide an energy consumption number that considers not only the computation, but, also, elements (e.g. links and devices) related to the communication.

In order to use standard metrics, we adopt concepts published in researches found in [1] and [2]. These researches take into consideration costs related to bit transmission from different type of users in a cloud configuration, and then indicate an appropriated configuration.

Therefore, our proposal considering those standard metrics is to evaluate the efficiency of a data center configuration utilized as a private cloud provider. The evaluation adopted in this paper comprised two phases. The first step was characterized by an empirical measurement and the calculation of energy consumption. In the second part of our research it was evaluated power consumption through the GreenCloud simulator. This procedure can appear in a reverse order, but the idea was to learn how to calculate the energy consumption and then scale the configuration size to a large environment.

The article is organized as follows. In second section some basic concepts related to green IT, characterization of devices energy consumption and energy efficiency are presented. Consumption metrics, calculus model and energy consumption analysis are illustrated in the third section. The fourth section shows the cloud configuration study case scenario, describing characteristics referring to the computational central core and corporate configuration. Experimental results and presented and discussed in the fifth section. Finally, in the sixth section we present our conclusions and research future directions.

2. Concepts Context

Concepts related to green cloud, energy efficiency and network devices utilized along this paper are explained in this section. Usually, those concepts can be employed broadly in several different contexts. Therefore, the goal of this section is to provide a primary idea on how these concepts are used in cloud computing scenarios.

2.1. Green Cloud

The recent trend on the intensive utilization of cloud configurations, as the paradigm to delivery services and execution of applications in a global fashion, brings also together a considerable increase on energy demand to support these data centre configurations.

The Gartner group [3] indicates that the data centres energy consumption reaches around 10% of operation costs of total costs. These costs are also known as *Operational Expenditure* (OPEX). Some studies (e.g. [1]) indicate that this energy consumption number can reach around 50% in years to come.

A study presented by the *Environmental Protection Agency* (EPA) [4] shows that the American data centres have energy consumption equal to 4.5 billion of kWh per year. This number represents 1.5% of energy consumption of the country. The research also indicates that the energy consumption double during the period from 2000 and 2006. Thus, it is expected that the new model of business based on cloud computing could lead the energy consumption to higher levels.

Heating is other important element which should be taken into the account of energy consumption in a cloud configuration, where it is related to infrastructure temperature maintained, because of machines of the environment. On other words, it is reported in commercial literature that the heating created by the machines from a cloud configuration can represent a cost between 2 and 5 million, considering a standard data centre. Therefore, as it is shown in [3], an optimized infrastructure can reduce costs related to the OPEX.

2.2. Energy Efficiency

As it is presented in [5] and [6], energy efficiency is traditionally a vital element to any kind of industry. The new type of industry in this scenario is the IT companies, because the energy consumption is been shown as relevant cost in their scope [12].

Following [7], for the cloud computing viewpoint the energy efficiency can be understood as:

- A set of computational resources and links organized in a way that the input energy can be efficiently transformed in calculus, or data transfer, to support users' requirements.

As pointed out in [8], usually, when the energy efficiency parameter is mentioned in a document, the computational power from servers and how they generate heating are main concerns. In a second moment, the parcel of cooling is the targeting of the concern. The amount of energy consumption has several factors, with a variety of reasons, but the low utilization of computational configurations is indicated as the most important factor.

2.3. Network Devices

Several studies (examples are [2] and [9]) indicate the low use of computational resources for long periods. It is estimated that these numbers can reach around 70% of idleness against 30% of full resource capacity use.

Therefore, the result is a large number of computational resources with a high inefficiency levels to provide service. Thus, this fact generates waste of energy, heating dissipation and the power necessary for cooling to those unused resources.

It is important to mention that several researches do not observe the power consumption from network devices. Nowadays, these elements are becoming more and more essentials, and be taken into consideration, because clustered configurations are the main computational core. In addition, several links exist with different complexities.

The research presented in [10], network devices (e.g. switches, routers and communication links) are regarded as a 15% of power consumption from data centres.

Similar to the computational resources, such as servers and terminals, network devices can also have their voltage and frequency reduced, under certain amount of demand, targeting to decrease the energy consumption.

3. Metrics, Calculus Model and Energy Consumption Analysis

In this section, it is presented some metrics, the calculus model and how is the process to verify and analyse the energy consumption, adopted in this research work.

3.1. Green Cloud

The general proposal of energy consumption metrics is to provide a global view of the infrastructure energy efficiency, especially, considering the consumption of element IT.

The era of cloud configurations has created a high demand for costs control from energy. This fact can be explained because data centres are growing in a form not known before. In addition, a lack of an official standard, or a *the facto* one, accepted by the community creates inconsistencies in practices to measure energy consumption.

Manufacturers, usually, have their own practice to measure metrics of consumption. Therefore, it is essential a recommendation to those organizations, that are seeking for a standard procedure to measure and analyse energy consumption from their IT infrastructure. In other words, providing a clear and objective method on how to measure and analyse energy consumption from a cloud configuration.

In this research work, we consider the *power usage effectiveness* and *data centre infrastructure efficiency* as metrics to measure cloud environments energy consumption. These two metrics are being accepted as reasonable parameters related to the data centres energy consumption.

3.1.1. *PUE.* The power usage effectiveness (PUE) is a parameter that seeks for efficiency utilization of the energy provided to a data center. This metric is recommended to characterize and provide information related to energy from the entire data center infrastructure [16].

PUE utilizes the energy consumption as a source for the calculus. PUE is characterized by a comparison between the total energy from a data center infrastructure and the IT energy parcel, thus:

PUE = (Total data centre energy consumption) /
(IT energy consumption)

3.1.2. *DCIE.* The *data center infrastructure efficiency* (DCIE) is characterized by the inverse calculus of the PUE. The DCIE goal is a complement to the PUE metric, because represents the efficiency percentage from the data center infrastructure.

DCIE = (IT energy consumption) /
(Total data centre energy consumption)

3.2. Calculus Model
The calculus model considered in this paper is based on the proposal presented in [2]. This proposal takes into consideration the bit energy as fundamental measure of consumption. Important to remember that, into a cloud configuration, the energy efficiency can be translated to a number that represents the consumption energy spent to execute a set of data bits, which were used through the cloud environment.

The E_c energy, following [2], can be expressed as:

$$Ec = 3\ x\ 3x \left(\frac{Ples}{Cles} + \frac{3Pes}{Ces} + \frac{Pg}{Cg}\right) (1)$$

Where, P_{les}, P_{es} e P_g correspond, respectively, to the energy consumption related small Ethernet switches, switches and routers.

On the other hand, the parameters C_{les}, C_{es} e C_g mean all the equipment capacity in bits per second.

The first 3 factor from *equation 1* is related to redundancy energy (this number is assumed as equal to 2). The calculus also takes into consideration the requirement for cooling and energy lost (this number is considered as equal to 1.5).

Usually, it is estimated that a private cloud configuration has the utilization around 33% of available time. Therefore, the second factor of the equation 1 is related to this figure from the environment. In other words, this factor considers energy consumption from the corporate switches, as well as the data centre's LAN.

3.3. Service Consumption Analysis
As it was mentioned in section 3.2, we adopted the proposed formula found in [2] for our service consumption analysis. The proposal states that the measure of energy consumption from a service, such as SaaS (*Software as a Service*), considers that a specific user is able to access an application in a cloud, where technical details are transparently.

Therefore, the energy consumption is summed up considering per usage from each user related to the terminal function, from where the service was required. Terminal can be understood as several devices that are shown in table 3, which can access services from a cloud configuration.

As stated in [2], the energy consumption from a user can be considered by the following equation:

$$P_{sf} = P_{sf,PC} + \frac{1.5P_{sf,SR}}{N_{sf,SR}} + 2B_d \frac{1.5P_{SD}}{B_{SD}} + AE_T \ (2)$$

Journal of Physics: Conference Series **540** (2014) 012005 doi:10.1088/1742-6596/540/1/012005

Where, $P_{sf,PC}$ is the energy consumption from a user terminal, whereas $P_{sf,SR}$ is the energy consumption from a server divided by the number of users that access this computational facility.

B_d (bits) refers to the average size of a file. Ordinaries documents and spread sheets, usually, a conventional number used is 1,25MB. Inside the equation the multiplication for two means the redundancy factor.

P_{sd} refers to the power consumption from disk storage and B_{SD} (bits) means the storage capability.

The A factor refers to data rate´s exchange between each user and servers. It is calculated by frames per second, because this service has a characteristic where computation and storage are executed by the cloud configuration. In other words, the terminal is used only as an output device.

$$A = 128\, x\, 1024\, x\, 24\, x\,\ Y\, b/s \quad (3)$$

As an example, it is possible to assume that the resolution of a terminal is equal to 128 x 1024, with a 24 bit color. The Y factor is the update rate from the terminal, which deserves the following explanation:

- It is considered a frame per second, the change of 100% of a user´s terminal per second. When only a part of the terminal received an update, only this part of the frame will be transmitted and accounted for the calculus.

E_t is the basic energy consumption per bit for data transmission and commutation on the network.

4. Cloud – Study Case

In this section, it is illustrated the cloud environment utilized as a study case in our research. The object is to explain details from the cloud configuration, shown in figure 1, used for our experiments. The study case configuration has two main components, a *corporate network* and *data center* environment.

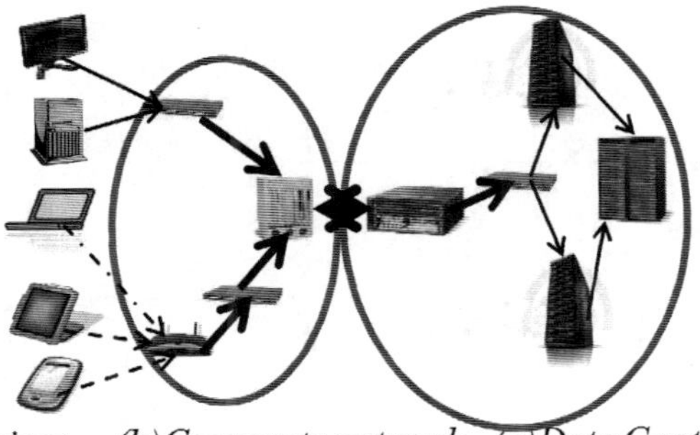

(a)Devices (b)Corporate network (c)Data Center

Figure 1: Cloud configuration for study case

4.1. *Corporate Network*

On the corporate side, available devices and power consumption are shown in tables 1 and 3. In table 1, it is possible to observe network devices used to access a cloud configuration (e.g. access and aggregation switches). On the other hand, table 2 shows examples of conventionally devices used to require services are personal computers, mobile telephones, tablets and laptops. All listed equipment will be included into the calculus of energy consumption.

4.2. *Data Center*

In this research work it is adopted a private cloud, which has a support from a small data centre. The configuration has architecture with two switches types (e.g. access and aggregation), two SMB servers and gateway router. This is a typical small cloud configuration.

In table 1, it is shown the BNG, PowerEdge, Core Router and WDM as network devices, from the data centre configuration. These devices are characterized from their manufacturer's specification with the bandwidth related to specific energy consumption.
Sections should be numbered as follows:

Table 1: Characteristics of network equipment

Network Device		**Capacity**	**Energy Consumption**
Switch	***Manufacturer***		
Access	HP/3COM4210	128 Gb/s	370W
Aggregation	HP/3COM 6600	176 Gb/s	261W
BNG	Juniper E320	60 Gb/s	3.3 kW
P. Edge	Cisco12816	160 Gb/s	4.21 kW
Core Router	Cisco CRS-1	640 Gb/s	10.9 kW
WDM800km	Fujitsu 7700	40 Gb/s	136 WChannel

On the hand, table 2 illustrates servers which are, usually, used inside the core of a data centre. Computational cores are responsible for the executions of applications. The performance to execute an application is related to a number of characteristics, such as floating point, memory access and an appropriated parallel execution. All the data from this table were obtained from manufacturers´ specification. There are several energy consumption software tools that may be used also to build a similar table. However, looking for to use a more accurate set of numbers we assume information from the manufacturers.

Table 2: Characteristics of servers from the cloud

Servers Configuration	**Processor Consumption**	**Device Consumption**	**Capacity**
	Idle Max.	**Total Load**	
HP Proliant Dl360 G5	120W 135W	337W	800 Mb/s
DellPower EdgeR710	80W 135W	376W	1,6 Gb/s
Storage Huawei		570W	5,86 TB

Table 3: Device available to access the cloud environment

Device Type	**Processor Consumption**	**Device Consumption**
	Idle Max.	**Total Load**
PC-Desktop	10W 65W	132W
Tablet	-	1-2W
Notebook	2.5W 35W	40W
Thin Client	8W 8W	-
Old PC	40W 103W	210W

5. Experimental Results

This section is characterized by two parts. In the first part, results from empirical experiments are presented. The use of the GreenCloud [11] simulator is utilized in the second part of the section. In addition, in the second part has some details related to the energy consumption. The objective, similar to [13],[14] and [15], was to learn how to calculate the energy consumption and then simulate computational and network elements from a specific configuration size.

5.1. *Empirical Study Cases*

Inside private clouds, where it is considered a universe of few users, it is possible to notice the high requirement of amount of energy necessary to transport data inside the configuration. This fact is especially true to the update rates from frames with high transmission rate.

Figure 2 shows energy consumption from several devices from our configuration, which represents a real cloud environment inside a large Brazilian organization. The figure it was generated by empirical results and from the calculus, using those formulas from section 3. The graphic presents a low energy consumption tendency when the number of users per server grows. On the other hand, the power consumption, related to transport (i.e. network devices) increases, to allow the greater flow.

Results presented in figure 2 represent the energy consumption per user in a private cloud with a universe of 20 users, with different devices and a comparison to an old fashioned computer.

The consumption of energy from the cloud can be considered low and stable, the threshold is characterized by 30W and 40W and the rate is 1 frame per second.

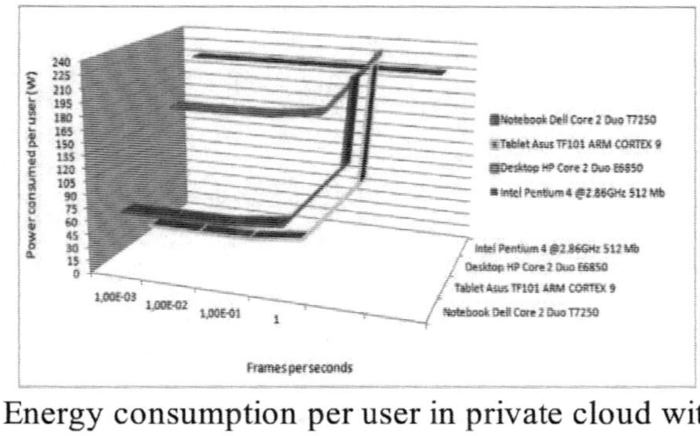

Figure 2: Energy consumption per user in private cloud with 20 users

The best case of energy consumption was verified by the tablet device, it is possible to verify that the required energy is very low, could be almost ignored for the calculus of power consumption. In contrast, the worst case was verified by desktop equipment which requires high rates of update. The figure from this equipment is similar to the old fashioned equipment, a Pentium 4, a machine which does not have any energy manager tool.

A private cloud has one advantage, in comparison to public cloud configuration, because it has a small network infrastructure. Thus, this element has a low weight for the power consumption calculus.

Figure 3 shows a study case of energy consumption per user in a private cloud with 200 users and a comparison with power consumption of a new notebook. It is possible to verify a sharp drop on energy consumption, where the threshold is located between 5W and 10W.

Results shown in figure 3 were led by the low energy consumption from infrastructure equipment to support the high number of users. These results indicate that cloud configurations are interesting environments for a large number of users that can reduce costs related to power consumption per user.

Another fact to be observed is the throughput from servers. In this study case, the upper throughput limit from the server was 800 Mb/s. Therefore, the graphics stopped at 0,12 frames/s, because this is the server limit to a configuration with 200 users.

In addition, it is interesting to observe that the energy consumption for the Dell notebook was equal in the case of direct access and to access the cloud configuration. However, this figure was not true for the access from small devices (e.g. tablets), where the energy consumption was different.

5.2. *GreenCloud*

In a second stage of our research work we considered the utilization of the GreenCloud simulator [11]. Table 4 shows all chosen parameters, these values were selected to provide a similar configuration to the designed cloud environment.

Analysing figure 4, it is possible to notice the high costs of energy related the network infrastructure for the small data centre. This fact reaffirms our assumption, which it is also stated in similar research works (e.g. [10]). This consumption number can represent circa from 10% to 20% of the data centre budget.

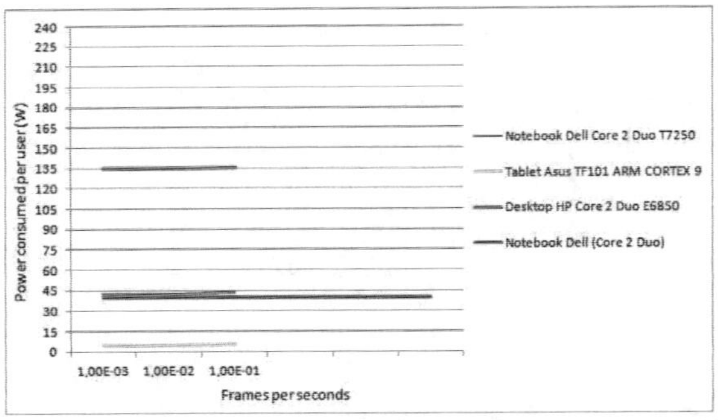

Figure 3: Energy consumption per user in private cloud with 200 users

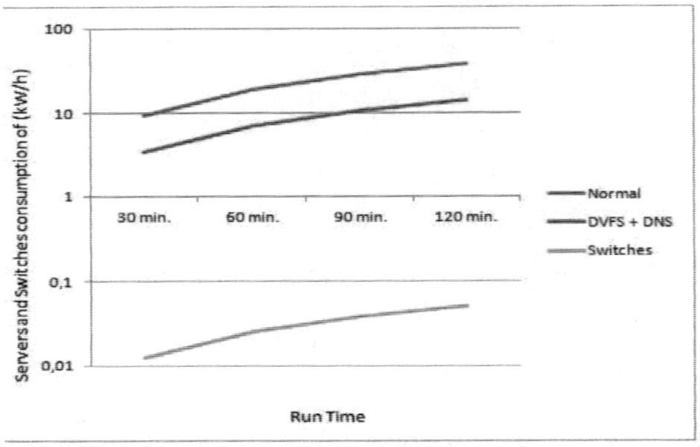

Figure 4: Servers/switches energy consumption for different execution time

There is an indication, when using the GreenCloud simulator package, of the influence in the use of the *Dynamic Voltage/Frequency Scaling* (DVFS) and the *Dynamic Shutdown* (DNS) related to power consumption from the devices. Thus, it is important to notice two factors:

- **Servers**: in the case of these devices there is a clear advantage in terms of energy management saving when the DVFS is used together with DNS. This utilization produce a workload balancing among servers, avoiding peak consumption, as it is shown in figures 5 and 6;
- **Network devices:** because our research work was based on a small private cloud configuration, with a small number of applications submission, we reached the 1W of power consumption. Adopting the use of DVFS and DNS for switches and routers, it was possible to reduce unnecessary energy consumption from ports and linecards. These two elements are responsible for the major energy consumption from network devices.

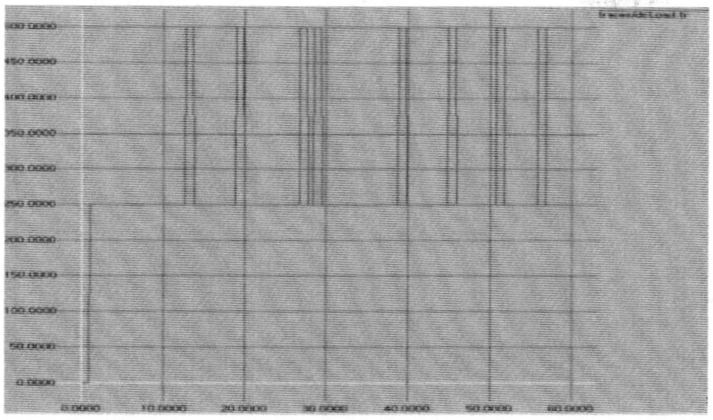

Figure 5: Workload without DVFS+DNS management

We analysed an energy bill from the real environment, which was used as a configuration for this study, without and with DVFS+DNS management. In this analysis it was considered factors such as IT, light, cooling, lost and others.

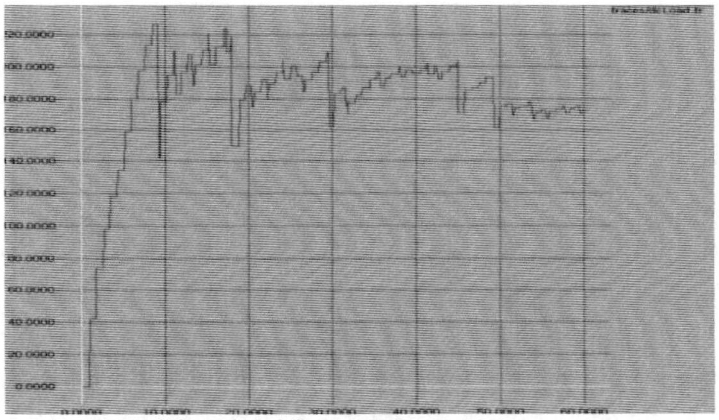

Figure 6: Workload with DVFS+DNS management

The PUE number, without energy management, was equal to a 1.63, where the total energy per month was 28,944 kWh and the consumption from the IT was 17,744 kWh. In contrast, the PUE number, with energy management, was equal to 1,76, where the total energy per month was 11,650 kWh and the consumption from IT was 6,600 kWh.

6. Conclusions and Future Work

This paper it was presented a research study case which had the goal to analyse the energy consumption from a cloud configuration, considering computational resources and network devices. Therefore, a corporate network and one small data centre were configured as a private cloud configuration. Measures from empirical experiments had shown that the data transport inside the configuration have an influence in the power consumption. Inside the private cloud configuration, our results indicated the efficiency of this environment in comparison to the utilization of standalone computers. In addition, the experimental results indicate the advantage of simultaneous utilization from several users. This fact can be explained by the enhanced distribution of cost of energy consumption through all users from the cloud. In addition, we used the GreenCloud package, where executing some simulations of the configuration, without and with energy management, pointed out an energy saving number of around 37%. On the other hand, the PUE calculus showed that some elements (e.g. lost fraction) should also take into consideration to provide a more accurate number of efficiency.

As a future work, we plan to utilize a set of scheduling algorithms to submit applications to a cloud environment, considering an energy consumption factor. Thus, it will provide a more green approach for the cloud configuration.

Table 4: Parameters utilized in the GreenCloud

Parameters	*Data centre Architecture*
	2 Layers
Switches	
Core	1
Aggregation	2
Access	2
Servers	4
Links	
Core - Aggregation	10 Gb
Aggregation Access	1Gb
Access - Servers	1Gb
	Others
Data centre load average	30%
Tasks size	4500 (3 Ethernet frames)
Simulation time	30, 60, 90, 120 minutes

References

[1] Kliazovich, D.; Bouvry, P.; Audzevich, Y.; Khan, S.U. GreenCloud: A Packet-Level Simulator of Energy-Aware Cloud Computing Data Centers, *GLOBECOM 2010, 2010 IEEE Global Telecommunications Conference*, pp.1-5, 6-10 Dec. 2010.

[2] Baliga, J.; Ayre, R.W.A.; Hinton, K.; Tucker, R.S.; Green Cloud Computing: Balancing Energy in Processing, Storage, and Transport Proceedings of the IEEE , vol.99, no.1, pp.149-167, Jan. 2011.

[3] Gartner Group http://www.gartner.com, available in June de 2012.

[4] EPA, http://www.epa.gov/, available in July 2012.

[5] Mahadevan, P.; Sharma, P.; Banerjee, S.; Ranganathan, P. Energy Aware Network Operations. IEEE INFOCOM, pp. 1 – 6, 2009.

[6] Mahadevan, P.; Sharma, P.; Banerjee, S.; Ranganathan, P.A Power Benchmarking Framework for Network Devices. IFIP-TC 6 Networking Conference, Aachen, Germany, May, 2009.

[7] Greenpeace. Make IT Green: Cloud Computing and its Contribution to Climate Change, Greenpeace International. March, 2010.

[8] Fan, W-D Weber, L. Barroso, Power Provisioning for a. Warehouse-sized Computer, Proceeding of ISCA'07, 2007.

[9] Dantas, M A R and Zaluska, E J (1998) Efficient Scheduling of MPI Applications on Networks of Workstations. *Future Generation Computer Systems*, 13, (6), 489-99.

[10] Greenberg, A.; Hamilton, J.; Maltz, D.; Patel, P. The cost of a cloud: Research problems in data center networks. In ACM SIGCOMM CCR: Editorial Note, Jan. 2009.

[11] GreenCloud, http://greencloud.gforge.uni.lu/, available in July 2012.

[12] Koomey, J. Estimating Total Power Consumption by Servers in the U.S. and the World. Oakland, CA: Analytics Press, 2007.

[13] Corradi, A.; Fanelli, M.; Foschini, L. , *Increasing Cloud power efficiency through consolidation techniques*, Proceedings of the IEEE Symposium on Computers and Communications (ISCC), pp. 129-134, 2011.

[14] Kantarci, B.; Mouftah, H.T. *Overcoming the energy versus delay trade-off in cloud network reconfiguration,* Proceedings of the IEEE Symposium on Computer and Communication (ISCC), pp. 53-58, 2012.

[15] Rogério Castellar Monteiro, M. A. R. Dantas, *A Preliminar Study of Energy Consumption in a Private Cloud*, Proceedings of Brazilian Symposium on High Performance Computing – Workshop of Scientific Initiation, Petropolis, Brazil, 2012.
[16] Green Grid, http://www.thegreengrid.org/events/tech-forum-2011.aspx, available February 2013.

Journal of Physics: Conference Series **540** (2014) 012006 doi:10.1088/1742-6596/540/1/012006

Efficient parallelization of molecular dynamics simulations with short-ranged forces

Ralf Meyer

Department of Mathematics and Computer Science and Department of Physics, Laurentian University, 935 Ramsey Lake Road, Sudbury, Ontario P3E 2C6, Canada

E-mail: `rmeyer@cs.laurentian.ca`

Abstract. Recently, an alternative strategy for the parallelization of molecular dynamics simulations with short-ranged forces has been proposed. In this work, this algorithm is tested on a variety of multi-core systems using three types of benchmark simulations. The results show that the new algorithm gives consistent speedups which are depending on the properties of the simulated system either comparable or superior to those obtained with spatial decomposition. Comparisons of the parallel speedup on different systems indicates that on multi-core machines the parallel efficiency of the method is mainly limited by memory access speed.

1. Introduction

Molecular dynamics (MD) is a particle based simulation technique that is widely used in many areas of computational science [1, 2, 3]. As computers became more and more powerful, simulations of larger and more complex systems became possible. Yet there are still many problems that can only be solved by even more powerful computers.

For the last years, increases of computing power were achieved through the integration of a growing number of processing units on so-called multi-core chips. In the near future processors with hundreds or thousands of CPU cores will become available. The power of such processors, can only be harnessed through highly parallel algorithms that achieve good parallel efficiencies for large numbers of threads.

Large-scale MD simulations in materials science frequently use short-ranged force models that are constructed in such a way that the force between two particles is zero if the distance between the particles exceeds a cutoff radius r_{cut}. A standard method for the parallelization of MD simulations with short-ranged forces is spatial decomposition [4]. In this method, the volume of the simulated system is subdivided into as many subvolumes as there are processors. Each processor then calculates the forces between particles in one of the subvolumes.

Spatial decomposition as a parallelization strategy for MD simulations has the advantage that it can be used on SMP machines as well as distributed systems using message passing techniques. The method scales well if the simulated system is large enough that the communication overhead can be neglected and sufficiently homogeneous to avoid load-balancing issues. The latter condition, however, becomes difficult to fulfill in simulations of complex nanosystems. If the simulated system contains a substantial amount of void volume or if multiple materials with different types of interactions are present, the even distribution of the computational workload among the processors becomes challenging.

In a recent article, a new algorithm for the parallelization of MD simulations on SMP systems has been proposed [5]. The *cell task* method uses task-based programming which is a modern approach to parallel programming that is particularly suited for irregular algorithms [6]. In this approach, the problem is divided into small work units (tasks) without regard for low-level details like the number of processors. A task-managing software then executes the tasks with the help of a thread pool. While a task is running it can create new tasks that are added to the list of tasks. If there are much more tasks then processors, a dynamic assignment of the tasks to the threads masks differences in the execution time of the tasks and keeps all processors busy.

In this article the efficiency of the cell task method is tested on a variety of different multi-core systems.

2. The cell task algorithm

The cell task method was designed with two goals in mind. First, the method should be highly parallel so that it performs well on many-core systems with possibly hundreds or thousands of threads. Second, the algorithm should provide similar speedups for systems of the same size independent of the homogeneity of the configuration.

The computationally most intensive part of any MD simulation is the calculation of the forces on the particles. When using short-ranged forces the algorithm for the force calculations has typically the following structure:

```
for all particles p:
    for all neighbors n of particle p:
        Calculate force F_{n-p} between particles n and p.
        Add F_{n-p} to total force on particle p.
        Add -F_{n-p} to total force on particle n.
    end for
end for
```

In this algorithm, the neighbors of a particle are other particles within the cutoff radius r_{cut}. For each particle the program maintains a list of its neighbors that is updated in regular intervals. In order to save time, the algorithm exploits Newton's third law which says that $\mathbf{F}_{n-p} = -\mathbf{F}_{p-n}$ and assumes that each pair of neighbors appears only in the neighbor list of one of the two particles.

On an SMP machine, the outer loop of the algorithm given above can easily be executed in parallel using for example OpenMP [7, 8]. Unfortunately, this simple strategy is not efficient. The reason for this is the need for synchronization of the force updates in the inner loop. This is necessary to avoid race conditions when multiple processors update the same particle simultaneously. In addition to this, uncoordinated accesses to the same particle by different processors can increase the amount of cross-socket transfers of cache lines.

Most programs for MD simulations with short-ranged forces employ the linked-cell technique [1] during the construction of the neighbor lists. The linked-cell technique subdivides the simulation cell into a grid of small cells and bins the particles into these cells. The size of the cells is chosen to be just larger than the cutoff radius of the interactions r_{cut}. The cell information can then be used to reduce the number of candidates for the neighbors of a particle. If a particle is in a given cell, its neighbors are all within the 27 immediately surrounding cells.

The cell task method reuses the information of the linked-cell technique to avoid synchronization constructs in the inner loop of the force calculation. If a processor calculates the forces for a particle p in the algorithm given above, it is known that all neighbors n will be found in the 27 cells surrounding the particle. Therefore, if it can be guaranteed that no other processor updates the particles in these 27 cells, no synchronization of the particle updates in

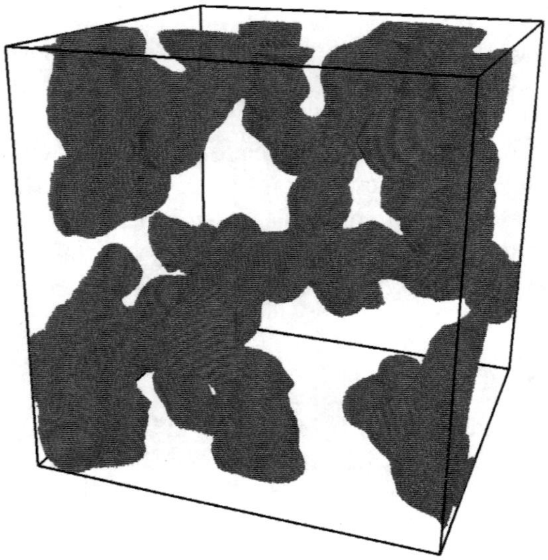

Figure 1. Configuration of the partially sintered nanocrystalline copper system

the inner loop is necessary. The other processors are however free to work on particles in other cells.

The cell task method organizes the force calculation into tasks. A task consists of the calculation of the forces on all particles within one grid cell (or a compact block of cells). The tasks are then organized in a dependent task schedule that ensures that simultaneously running tasks will only access non-overlapping sets of cells. A similar scheme is employed for the generation of the neighbor lists.

The current implementation of the cell task method is build on top of a general purpose MD program named `mdntp` [11]. This program has been designed for large scale simulations using many-body interactions like embedded-atom method [12] or Finnis-Sinclair style potentials [13]. The cell task version of `mdntp` uses Intel's *Thread Building Blocks* Library [9, 10] for the management of the tasks. Details of the construction of the task schedule and the implementation of the method can be found in [5].

3. Results from benchmark simulations

3.1. Benchmark Systems

In order to test the efficiency of the cell task method, three types of benchmark simulations have been repeated on a variety of multi-core systems. All simulations employed Cleri and Rosato's tight-binding second-moment potential for copper [14].

The first benchmark configuration is a cubic block of crystalline bulk copper build from $63 \times 63 \times 63$ fcc unit cells (1,000,188 atoms). This system is very homogeneous and therefore nearly ideal for the spatial decomposition method. Load imbalances occur in this system only if the spatial decomposition is incommensurable with the number of lattice planes.

The second benchmark system is a spherical copper nanoparticle with a diameter of 30 nm containing 1,177,151 atoms. Since this system does not fill the space completely, spatial decomposition becomes problematic if the decomposition does not cut the system into equal parts. Good performance of the spatial decomposition method can be expected for this system only for 2, 4, and 8 processors where an efficient decomposition is possible by cutting the sphere repeatedly in halves.

The last benchmark configuration is a porous system of nanocrystalline copper that was obtained from simulations of the sintering of copper nanoparticles (1,992,220 atoms; see figure 1). This system is very inhomogeneous with an irregular shape and a large fraction of void volume.

A good load balance can not be expected from a simple decomposition of this system.

All simulations were repeated five times and the average execution time (excluding initialization and saving of results at the end) was calculated from the five runs. The execution times were then used to calculate the parallel speedup with respect to the serial version.

Nearly direct comparisons between the parallel efficiency of spatial decomposition and the cell task method are possible since the original version of mdntp supports spatial decomposition. However the implementation of the cell task method in the program required some changes to the force calculation that lead to additional optimizations that are not present in the original version and that can not easily be imported into the spatial decomposition version of the program. For this reason two different serial programs were used for the calculations of parallel speedups. Parallel speedups $S(p)$ of the spatial decomposition method were calculated with respect to the execution time of the original serial program version. Speedups of the cell task method were calculated by dividing the execution time of the cell task version of the program running with 1 thread by the execution time of the same program running with p threads. This procedure effectively eliminates the effect of the additional optimizations that have nothing to do with the parallelization method.

3.2. Comparison with spatial decomposition

Figure 2 shows the speedups obtained for the three benchmarks system with the cell task method and spatial decomposition on three different multi-core machines. The figure shows that except for very few points in panel (g) the cell task method obtains better parallel speedups than the spatial decomposition method. As expected, the two methods give similar speedups in case of the homogeneous bulk system (left column). For the other two benchmark configurations the cell task method performs significantly better.

It can also be seen from figure 2 that the cell task algorithm yields considerably more consistent speedups. The speedups obtained by the cell task method increase continuously with the number of threads and on the same machine the behavior of the three benchmark systems is rather similar. In contrast to this the speedup of the spatial decomposition method is non-monotonous and the results vary from one benchmark system to another on the same machine.

Finally, figure 2 shows a growing deviation of the speedup from the ideal behavior. It is however unlikely that this loss of efficiency is due to a lack of parallelism of the cell task algorithm. If this were the reason, the spatial decomposition method should outperform the cell task method at least in the case of the bulk system since spatial decomposition scales well for large homogeneous systems. The left column of figure 2 shows however that for the bulk system the spatial decomposition method deviates from the ideal behavior in the same manner as the cell task method. This is a strong indication that the deviations are not caused by the parallel algorithms but by limitations of the test machines. In addition to this, benchmark data given in reference [5] show that the cell task method achieves parallel efficiencies above 80 % at $p = 60$ threads on a Xeon Phi coprocessor. The most likely reasons for the lower parallel efficiency on the multi-core machines at high numbers of threads are limitations of the memory system. This will be discussed in the next section.

3.3. Comparison between test systems

Table 1 presents for a variety of multi-core machines the execution times $t(1)$ of the cell task program with one thread and the speedups obtained with $p = 6$, 12, 16, 24 and 32 threads for the third benchmark system (porous nanocrystalline copper). A surprising observation is that despite the strong variations between the test machines, the speedup data are rather similar. Unfortunately most of the systems differ in more than one parameter. This makes it difficult to identify the parameters with the largest impact on the parallel performance.

Journal of Physics: Conference Series **540** (2014) 012006	doi:10.1088/1742-6596/540/1/012006

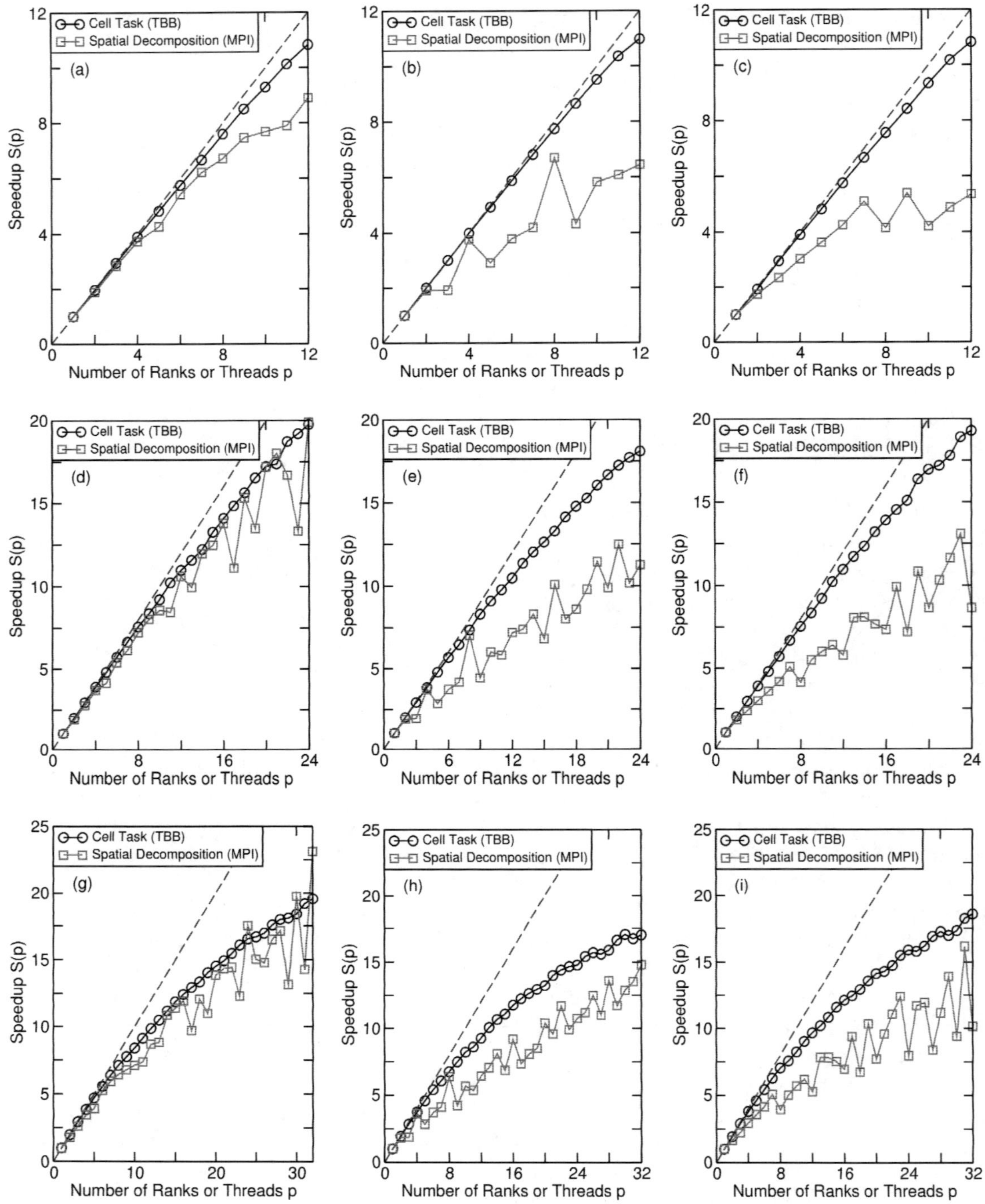

Figure 2. Parallel speedups obtained by the cell task method and spatial decomposition for the three benchmark systems bulk copper (left), copper nanoparticle (center) and porous nanocrystalline copper (right) on three test machines 12 core Xeon E5-2630 (top), 24 core Opteron 6174 (middle) and 32 core Opteron 8354 (bottom). Dashed blue lines indicate the ideal speedup of 1 per processor.

Table 1. Serial execution times $t(1)$ and speedups $S(p)$ for $p = 6$, 12, 16, 24 and 32 threads of the cell task method for the porous nanocrystalline benchmark system.

CPU Family	Xeon	Xeon	Xeon	Opteron	Opteron	Opteron	Xeon
CPU Model	X-5650	E5-2630	E-7340	6174	6238	8354	E5-4620
Cores	2×6	2×6	4×4	2×12	2×12	8×4	4×8
Clock Freq. [GHz]	2.7	2.3	2.4	2.2	2.6	2.2	2.2
Memory	DDR3 1333	DDR3 1600	DDR2 667	DDR3 667	DDR3 1600	DDR2 533	DDR3 1067
t(1) [s]	1030.2	1150.2	1772.1	1730.3	1305.3	1891.7	1020.4
S(6)	5.3	5.8	5.6	5.7	5.7	5.4	5.0
S(12)	9.7	10.8	10.4	10.9	10.5	9.7	9.3
S(16)			13.4	13.9	12.6	12.1	11.7
S(24)				19.2	15.8	15.9	15.4
S(32)						18.6	18.2

The two most similar systems in table 1 are probably the two Xeon E5 machines. The CPUs of these two systems have the same microarchitecture and very similar clock frequencies. Comparison of the two systems shows that better speedups are obtained on the system with faster memory. Another trend indicated by table 1 is that a low serial execution speed favors higher speedups at larger numbers of threads. This too is consistent with the idea that the parallel efficiency is mainly limited by the memory system since a slower program execution reduces the impact of memory access speed.

4. Summary and conclusions

The cell task algorithm provides an alternative parallelization method for MD simulations on multi- and many-core systems. The method is designed to be highly parallel and to give high speedups even for strongly inhomogeneous systems where spatial decomposition becomes less effective.

Benchmark calculations using three types of configurations on a variety of systems show that for inhomogeneous systems the cell task method yields a better parallel efficiency than spatial decomposition. For homogeneous systems the speedups are comparable.

On the multi-core systems used in this work, the parallel efficiency of the cell task method diminishes for larger numbers of cores. The reason for this is not yet fully understood. The fact that the spatial decomposition method is affected in a similar manner points however in the direction of system limitations rather than insufficient parallelism.

Although the data in table 1 do not allow to draw definite conclusions, they are compatible with the hypothesis that the main factor limiting the parallel efficiency of the MD simulations on the multi-core machines is memory access speed. More work is required to verify this hypothesis. One way to do this would be to perform simulations with a computationally more intensive potential. If the parallel efficiency in the current work is indeed bound by memory access speed, higher efficiencies can be expected for a potential that requires more computations per particle pair.

Acknowledgments
I wish to thank H. Merz from SHARCNET for his help to obtain the technical details of some of the computers used for the benchmarks. Financial support by Laurentian University and the Natural Science and Engineering Council of Canada (NSERC) is gratefully acknowledged. This work was made possible by the facilities of the Shared Hierarchical Academic Research Computing Network (SHARCNET:www.sharcnet.ca) and Compute/Calcul Canada.

References
[1] Allen M P and Tildesley D J 1987 *Computer Simulations of Liquids* (Oxford: Clarendon)
[2] Frenkel D and Smit B 2002 *Understanding Molecular Simulation* (San Diego: Academic)
[3] Rapaport D C 2004 *The Art of Molecular Dynamics Simulation* (Cambridge: Cambridge University Press)
[4] Plimpton S 1995 *J. Comp. Phys.* **117** 1–19
[5] Meyer R 2013 (*Preprint* `arXiv:1305.4196`)
[6] Korch M and Rauber T 2004 *Concurrency Computat.: Pract. Exper.* **16** 1–47
[7] `http://www.openmp.org/`
[8] Chapman B, Jost G and van der Pas R 2008 *Using OpenMP* (Cambridge: MIT Press)
[9] `http://threadingbuildingblocks.org/`
[10] James R 2007 *Intel Threading Building Blocks: Outfitting C++ For Multi-Core Processor Parallelism* (Sebastopol: O'Reilly Media)
[11] Meyer R 1998 Ph.D. thesis Gerhard-Mercator-Universität Duisburg Germany
[12] Daw M S and Baskes M I 1984 *Phys. Rev.* B **29** 6443–6453
[13] Finnis M W and Sinclair J E 1984 *Phil. Mag.* A **50** 45
[14] Cleri F and Rosato V 1993 *Phys. Rev.* B **48** 22–33

Journal of Physics: Conference Series **540** (2014) 012007 doi:10.1088/1742-6596/540/1/012007

High-Dimensional Adaptive Particle Swarm Optimization on Heterogeneous Systems

M P Wachowiak[1], **B B Sarlo, and A E Lambe Foster**
Department of Computer Science and Mathematics
Nipissing University, North Bay, ON Canada, P1B 8L7

E-mail: markw@nipissingu.ca

Abstract. Much work has recently been reported in parallel GPU-based particle swarm optimization (PSO). Motivated by the encouraging results of these investigations, while also recognizing the limitations of GPU-based methods for big problems using a large amount of data, this paper explores the efficacy of employing other types of parallel hardware for PSO. Most commodity systems feature a variety of architectures whose high-performance capabilities can be exploited. In this paper, high-dimensional problems and those that employ a large amount of external data are explored within the context of heterogeneous systems. Large problems are decomposed into constituent components, and analyses are undertaken of which components would benefit from multi-core or GPU parallelism. The current study therefore provides another demonstration that "supercomputing on a budget" is possible when subtasks of large problems are run on hardware most suited to these tasks. Experimental results show that large speedups can be achieved on high dimensional, data-intensive problems. Cost functions must first be analysed for parallelization opportunities, and assigned hardware based on the particular task.

1. Introduction

The literature on applying graphics processing units (GPUs) to Particle Swarm Optimization (PSO) is increasing. This is not surprising, as PSO is a relatively simple technique that has shown remarkable potential in solving a wide variety of problems in global optimization. PSO is a population-based stochastic global optimization algorithm that can optimize a wide range of cost functions, including large, complex ones [1, 2]. It simulates cooperative behaviour, as opposed to the competitive genetic algorithm and evolutionary strategy paradigms, and has motivated a great deal of research [3].

The inherent parallelism of PSO can be mapped onto high performance computing (HPC) systems, including distributed clusters [4] and, more recently, onto graphics processing units [5, 6]. Many papers have recently appeared that employ GPUs for PSO [7, 8]. Originally designed for 3D graphics rendering, GPUs are also ideal parallel processors for fine-grained, data-parallel, high-throughput and streaming applications, and exceptionally large speedup can be achieved when computations can be streamed onto them [5]. However, many very high-dimensional complex cost functions, such as those used in "big-data" applications, require expensive linear algebra operations, and often require large external data, such as transformation matrices, or large sets of empirical data. Due to the heavy computation required to evaluate these cost functions, the limited amount of memory available on

[1] To whom any correspondence should be addressed.

GPUs, and CPU-to-GPU data transfer bottlenecks, some combination of CPU and GPU processing would offer potential advantages.

Multicore technology for task-parallel medium-granularity tasks where fewer processing elements perform more intense computations also offers efficiency gains [9]. Due to limited memory bandwidth and inefficient memory management schemes, multicore processing at present offers only limited speedups [10]. Consequently, many-core (tens or hundreds of cores), digital signal processors (DSPs), acceleration processing units (APUs), field-programmable gate arrays (FPGAs) and other accelerators will likely lead to great performance gains in the near future [11]. Comparisons of GPU and multicore performance in PSO are also being investigated [6].

GPU and multicore HPC can be used in tandem, wherein each component works on tasks to which it is best suited. These heterogeneous systems are progressively gaining a foothold in many scientific and engineering applications (e.g. [12]). Heterogeneity is achieved by multicore parallelism complemented with accelerators (e.g. GPUs, Cells, DSP chips). This approach is taken in the current paper to increase the efficiency of high-dimensional and difficult global optimization of complex cost functions, wherein parallelization is achieved according to Schnabel's taxonomy [13]. Highly data-parallel but computationally simple tasks were accelerated with GPUs, while more intensive calculations were made more efficient through multicore processing. Assigning tasks to the different types of processors, while intuitive, was verified by profiling analysis. This paper demonstrates that implementing a heterogeneous, problem-specific approach can achieve good performance gains, and address the shortcomings of using strictly multicore (not scalable, not ideal for very fine-grained tasks) or GPU (not ideal for long computations requiring large data).

2. Adaptive Particle Swarm

The guiding principle of PSO is updating the D-dimensional positions of N independent "particles" that explore the D-D space in search of an optimum solution. These updates are based upon the progress of the search. Each particle's velocity update at time $t + 1$ is updated as:

$$\underbrace{\mathbf{v}_i(t+1)}_{\text{Updated velocity for particle } i} = \underbrace{\omega(t)\mathbf{v}_i(t)}_{\substack{\text{Inertia, affected by}\\ \text{previous velocity}}} + \underbrace{\overbrace{C_1\varphi_1}^{\substack{\text{Local acceleration}\\ \text{factor}\times\varphi_1\sim U(0,1)}}\left(\mathbf{p}_i^{\text{best}}(t)-\mathbf{x}_i(t)\right)}_{\substack{\text{Effect of particle } i\text{'s best position}\\ \text{found (local effect)}}} + \underbrace{\overbrace{C_2\varphi_2}^{\substack{\text{Global acceleration}\\ \text{factor}\times\varphi_2\sim U(0,1)}}\left(\mathbf{g}^{\text{best}}(t)-\mathbf{x}_i(t)\right)}_{\substack{\text{Effect of the globally best particle with}\\ \text{the best function evaluation}}} \quad (1)$$

Stochasticity, provided by random factors φ_1, φ_2 ~ $U(0, 1)$, maintains diversity in the population. After the new velocity is computed, the position of the i-th particle is updated as:

$$\mathbf{x}_i(t+1) = \mathbf{x}_i(t) + \mathbf{v}_i(t+1). \quad (2)$$

Many variations of particle swarm have been proposed since its inception. Such variations include hybridization with genetic and evolutionary operators (e.g. [14]), subpopulations [15; 16], hybridization with other optimization methods [17], and adapting various components of the PSO algorithm itself (e.g. [18]). A particularly promising approach is to adjust the inertial factor $\omega_t = \omega(t)$ in Eq. 1 (see [19] for a discussion of these methods and references). In the current paper, a specific adaptive PSO (henceforth, APSO) is considered, in which the inertial factor is related to particle clustering [20]. Specifically, in each iteration t, the inertial factor is defined by:

$$\omega_t(\phi) = \frac{1}{1+1.5e^{-2.6f}}, \; \phi = \frac{d_g - d_{\min}}{d_{\max} - d_{\min}}, \text{ and } d_i = \frac{1}{N-1}\sum_{j=1, j\neq i}^{N}\sqrt{\sum_{k=1}^{D}\left(x_i^k - x_j^k\right)^2}. \quad (3)$$

Here, the d_i represent the mean distance of particle $\mathbf{x}_i$ to every other particle (d_g for the globally best particle), requiring a distance matrix to be computed. Also, $d_{\min} = \min(d_i)$, $d_{\max} = \max(d_i)$, $i = 1, \ldots, N$, and ϕ is an evolutionary factor calculated to determine the state of the search based on particle clustering. In this formulation, $\phi \in [0, 1]$ so that $\omega(\phi) \in [0.4, 0.9)$. APSO further uses ϕ in a fuzzy membership function to determine the current state of the search, with small values ($\phi < 0.3$) generally indicating a convergent state of the swarm, followed by exploitation ($0.2 < \phi < 0.6$), exploration with more global searching ($0.4 < \phi < 0.8$), and large values ($\phi > 0.7$) indicating that the swarm needs to escape from local optima. The acceleration coefficients C_1 (local) and C_2 (global) are also modified on this basis [20]. In the convergence state, "elitist learning", where random dimensions of the globally best particle are perturbed, is also employed to discourage premature convergence. APSO has shown promise in both unimodal and multimodal problems, with adaptations based on particle positions and current search state, rather than on the basis of time (iteration) alone.

3. Methods

In his seminal paper, Schnabel [13] lists three potential levels of parallelism in global optimization; parallelization of: (1) the cost function; (2) any linear algebra operations, and; (3) the optimization algorithm itself. Schnabel focused primarily on multiple-instruction-multiple-data (MIMD) parallelism, indicating that single-instruction-multiple-data (SIMD) architectures are insufficiently general due to conditionals, branching, etc. in the cost function. Although SIMD systems have improved markedly since 1995, GPU-exclusive methods may not be conducive to evaluating many expensive cost functions. Consequently, in this paper, cost function evaluation is primarily assigned to CPUs, with SIMD (e.g. GPUs) used for linear algebra and high-throughput operations [13]. All multicore parallelism was achieved with OpenMP.

3.1. Cost functions

Two cost functions, one 200D composite and one realistic (but simplified) problem from geophysics, were optimized with heterogeneous APSO. Ground truth values for these functions were known.

3.1.1. Composition function (200D). Composition functions, weighted sums of transformed search spaces for standard test functions (e.g. Rastrigin, Weierstrass, etc. [20]) were proposed to test PSO in highly irregular, nonconvex space. These functions are characterized by multiple optima, and are extremely difficult to optimize, except in very low dimensions ($D < 10$). The general definition for these functions, $f_{comp}(\cdot)$, is given as [21]:

$$f_{comp}(\mathbf{x}) = f_{bias} + \sum_{m=1}^{M} w_m \left[f'_m\left(\mathbf{T}_m \left(\frac{\mathbf{x} - \mathbf{o}_m}{\lambda_m} \right) \right) \right], \text{ where } w_m = \exp\left(-\frac{\sum_{k=1}^{D} (x_d - o_{m,d})^2}{2D\sigma_m^2} \right). \quad (4)$$

Here, M denotes the number of individual functions in the composition. Each weight is computed for each individual function and for each particle on the basis of a standard deviation σ_m for each of the M functions, and $f'_m(\cdot)$, $m = 1, \ldots, M$, is $f_m(\cdot)$ scaled by the $\max[f_m(\cdot)]$ within the search space and by a constant, $\mathbf{o}_m$ is a vector of offsets of length D, $\mathbf{T}_m$ is a dense $D \times D$ orthogonal matrix, and λ_m is a scaling constant for each individual function. In this paper, $D = 200$, and $M = 5$. The individual cost functions are: {Rastrigin, Weierstrass, Ackley, Griewank, Weierstrass} (see [19, 21] for definitions). Because of the dense linear transformation in 200D, the search space is very complex, and therefore the goal is to determine a near-optimal solution, or at least a very good local optimum.

3.1.2. Geostatic correction (20D). A synthetic 20D geophysics optimization problem was implemented. In this problem, known as geostatic correction (GSC), subsurface images of geologic data are constructed from seismic reflection surveys. Because of noise or differences in surface materials, the images may become distorted, which may be corrected by shifts ("static corrections"). This problem is very difficult due to its high degree of nonlinearity and the presence of many local optima [22]. Genetic, evolutionary, and other approaches have been previously employed [22; 23]. The situation may be formulated mathematically as (see [22] for a more thorough description):

$$
\begin{aligned}
&A_1 = B_{r1,c1} - x_{c1} \\
&A_2 = B_{r2,c2} - x_{c2} \\
&p_{r1,r2,c1,c2} = \begin{cases} (A_1 + O_{r1,c1}) - (A_2 - O_{r2,c2}), & A_1 \le A_2 \\ (A_2 + O_{r2,c2}) - (A_1 - O_{r1,c1}), & \text{otherwise} \end{cases} \\
&f(\mathbf{x}) = -\sum_{c1=0}^{D-2} \sum_{c2=c1}^{D-1} \sum_{r1=0}^{Z-1} \sum_{r2=0}^{Z-1} p_{r1,r2,c1,c2}
\end{aligned} \tag{5}
$$

Here, a matrix of base values, $\mathbf{B}_{55\times20}$ consists of 20 plates (the number of dimensions) at 55 levels deep, and an offset matrix, $\mathbf{O}_{55\times20}$.

3.2. Parallelization of APSO

The entire APSO system was analysed to determine the most computationally intensive components, and the potential parallelism of each component was assessed. A profiling tool, *AMD CodeAnalyst*, collects samples uniformly as the program executes [24]. From the profiling results of the 200D composition with N = 8192, the most time-intensive components were the distance matrix (about 57%), matrix multiplication (about 16%), and the Weierstrass evaluation (about 9%), followed by the particle and weight update sections (each less than 1%). In the 20D geostatic correction problem, the largest component was the cost function itself (about 99%), followed by the distance matrix (about 1%). It was therefore decided to investigate parallelization of: (1) particle updates (Eqs. 1-2); (2) weight updates (Eq. 4); (3) the Weierstrass component to the composition cost functions (Eq. 4); (4) matrix multiplication (Eq. 4), (5) distance matrix calculation (Eq. 3), and (6) geostatic correction cost function (Eq. 5). These components are now described using Schnabel's taxonomy.

3.2.1. Parallelization of the cost function. The composition function is parallelized through the matrix multiplication (see below), and, because of the number of conditionals required, is not a candidate for further parallelization. However, the Weierstrass component of these compositions requires a moderate amount of computation, and was therefore parallelized. The geostatic correction problem is a straightforward four-level nested loop, and was therefore analyzed for parallelization.

3.2.2. Parallelization of linear algebra operations. Each evaluation of the composition function requires search space rotation (the dense $\mathbf{T}_m$ matrix in Eq. 4). The NVIDIA CUDA framework (used in this paper) supplies a linear algebra toolkit, cuBLAS, implementing BLAS routines optimized for the GPU. As a SIMD problem, matrix multiplication is an ideal task for the GPU.

3.2.3. Parallelization of the algorithm (distance matrix, weights updates, and particle updates). Calculating the d_i in Eq. 3 requires a distance matrix (DM). In [20], a small number of particles (N = 20) was used, and, consequently, the calculation was not considered to be intensive. However, forming a DM has $O(N^2)$ complexity, with $(N^2 - N)/2$ distances to be calculated, and can be intensive for large N. The DM can be parallelized via multicore by indexing the upper-triangular matrix. A

GPU-based method [25] was also implemented. Timing results for parallel methods were compared to the two-level nested loop method that directly computes the upper-triangular DM efficiently on one core, but, because of the load imbalance, cannot be easily parallelized. Weight and particle updates are straightforward loop parallelizations.

3.3. CUDA implementation
GPU kernels (routines that run on the GPU) were developed in the NVIDIA CUDA (Compute Unified Device Architecture) framework for GPU programming. CUDA also simplifies the transfer of information between RAM and the GPU-specific RAM. Parallelization is based on a number assigned to each thread. OpenCL, a specification from the Khronos Group, is an alternative framework that supports a wide variety of devices, including GPUs, multicore, and FPGAs, and has also been used for PSO parallelization by other researchers (e.g. [6]).

3.4. Experiments
Experiments were performed with populations of sizes N = 1024, 2048, 4096, and 8192 particles. For the component trials described below, dimensions of D = 16, 128, and 1024 were analysed. APSO trials were run on a computing cluster (SHARCNET) with characteristics of a commodity PC, as only one server unit was employed. Each unit contains two quad-core CPUs (Intel E5430) running at 3.0 GHz with 8 GB RAM, connected to 2 GPU servers, each connected to one NVIDIA Tesla S1070 graphics card with 4 GB GPU memory.

3.4.1. Speedup of APSO components. Speedup results for each individual parallelized APSO component described in Sections 3.2.1-3.2.3 were obtained as the mean of 100 tests. On the GPU, the DM was computed with an efficient algorithm [25] and reduced on the CPU, as the average distances (Eq. 3) were much more efficiently computed this way as opposed to GPU reduction.

3.4.2. Speedup of APSO experiments. Speedup results for full APSO trials for the two cost functions were obtained for the following configurations: (1) multicore, with no GPU acceleration; (2) composition function, multicore parallelization (including DM), cuBLAS acceleration for the search space transformation; (3) multicore parallelization with GPU acceleration of the DM computation.

3.4.3. Timing results of APSO experiments. Computation times for 500-iteration trials were obtained for: (1) no parallelization (1 core for all operations), and; (2) eight cores, GPU-accelerated DM calculation, and GPU matrix multiplication (composition function).

4. Results

Both cost functions were successfully optimized by APSO (figure 1). Because of the complexity of search space of the 200D composition function, a larger population size results in a better global solution. The 20D GSC problem was successfully optimized with only 1024 particles, but, with more realistic scenarios, the more complex cost function would benefit from larger N.

The individual APSO components were timed with 1, 2, 4, and 8 cores, and analysed for population sizes of N = 1024, 2048, 4096, and 8192, and dimensions of D = 16, 128, and 1024. Speedup results are shown in figures 2a-e, optimized cuBLAS matrix multiplication (composition) speedups are shown in figure 2f, and multicore DM calculations in figure 2g.

The full APSO algorithm with CPU-based DM computation was run with the 200D composition function with cuBLAS-accelerated rotational matrix multiplication (figure 3a). The 20D geostatic correction function was run with the CPU-based DM technique (figure 3b). Finally, both cost functions were run with the GPU-accelerated DM (and the GPU-accelerated matrix multiplication for the composition function) (figures 3c-d). Timing results for two configurations – one core, no parallelism vs. eight cores with GPU acceleration, are shown in Table 1. For the composition, speedup is shown for (1) 8-core with cuBLAS on the GPU, and (2) 8-core with cuBLAS and DM on the GPU.

For GSC, speedup is shown for (1) 8-core, and (2) 8-core with DM on the GPU. As their speedups and percentage of total computation were both low, weight updates were not parallelized.

From these results, it is seen that the heterogeneous approach achieves better speedup than multicore alone for the composition function. Because of their sizes, and given current limited GPU memory and CPU-GPU bandwidth bottlenecks, these problems are not runnable exclusively on the GPU, and, likely, other accelerators as well. GSC optimized faster with multicore (the problem is too large to fit all on the GPU). Therefore, for large problems, such as those arising in "big data" applications, analysis of the various optimization components, per Schnabel, leads to heterogeneous solutions that achieve acceptable levels of performance and efficiency gains.

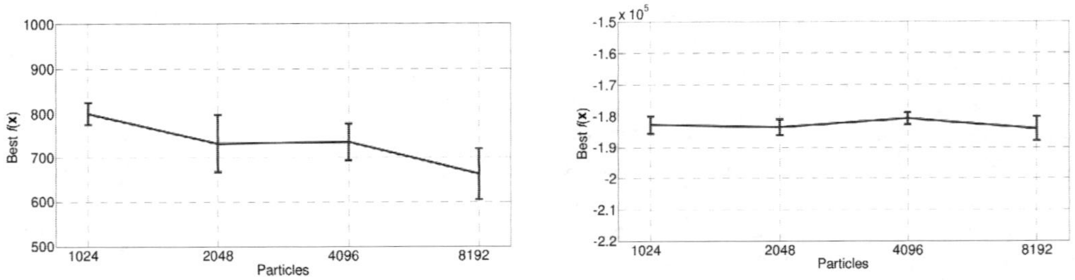

(a) 200D composition function. (b) 20D geostatic correction problem.

Figure 1. Best $f(\mathbf{x})$ cost function values returned by the globally best particle with varying N.

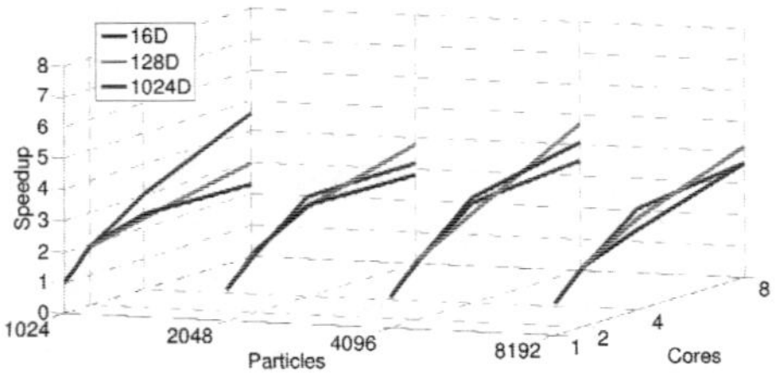

(a) Particle update.

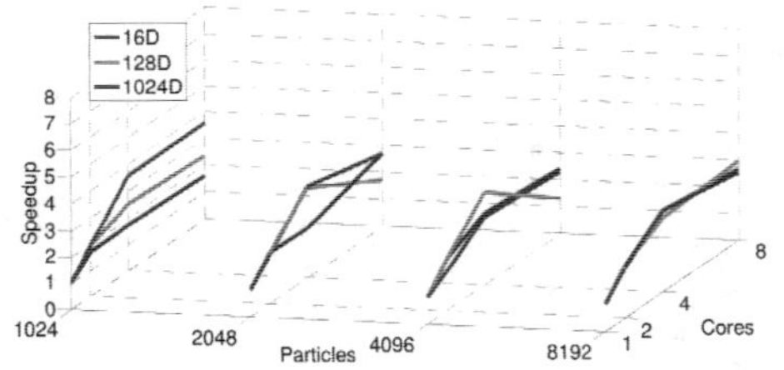

(b) Weight calculations (composition function).

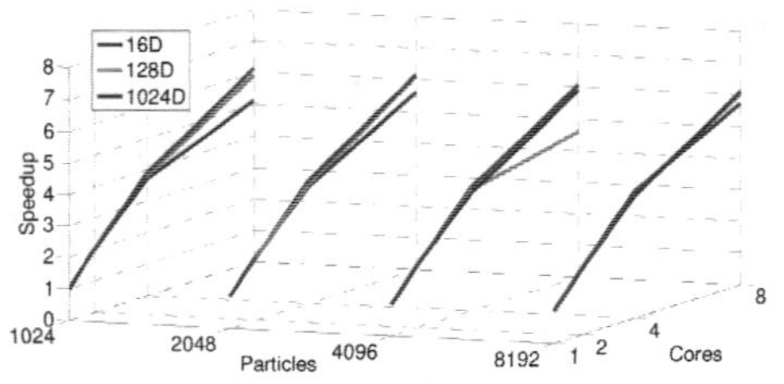

(c) Weierstrass function.

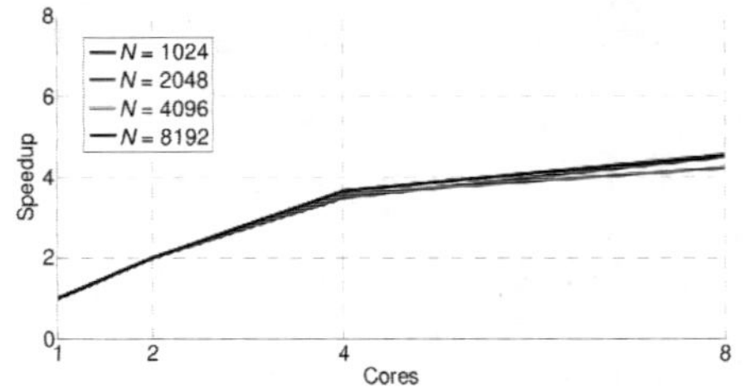

(d) Geostatic correction cost function.

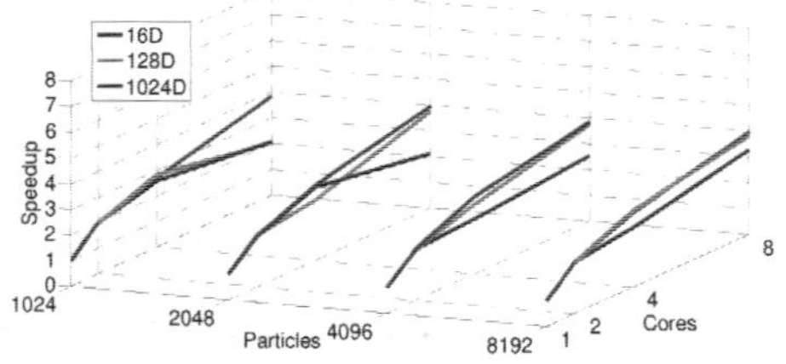

(e) Multicore matrix multiplication.

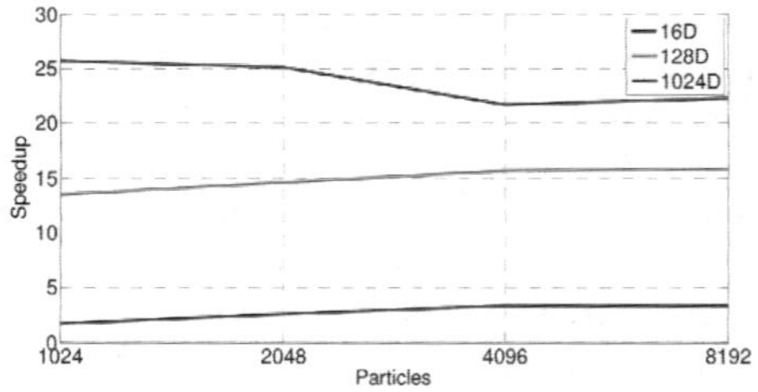

(f) cuBLAS matrix multiplication.

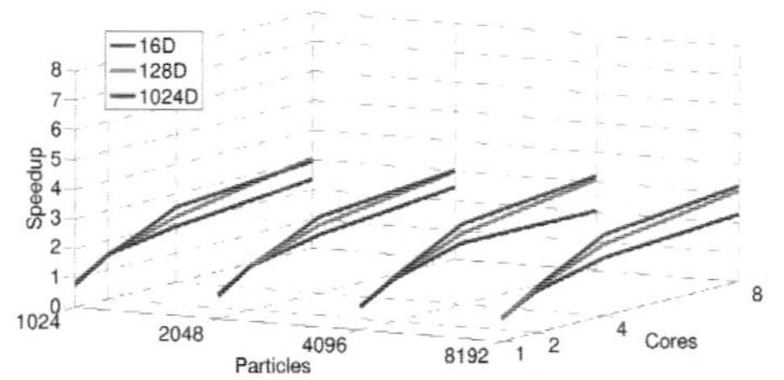

(g) Multicore distance matrix calculation.

Figure 2. Speedups for individual APSO components with varying N and D.

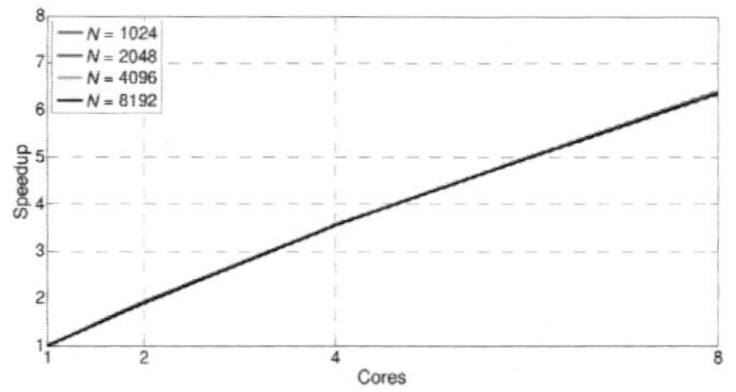

(a) Multicore, cuBLAS, 200D composition.

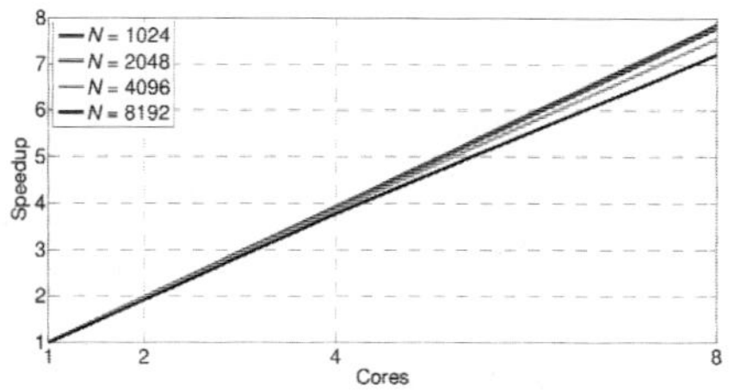

(b) Multicore, 20D GSC.

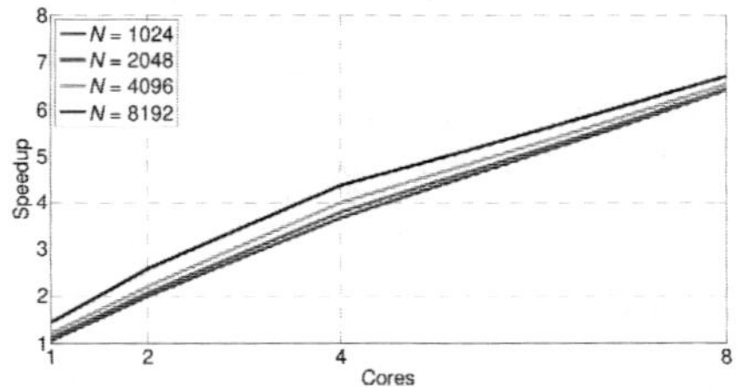

(c) Multicore, cuBLAS GPU DM, composition.

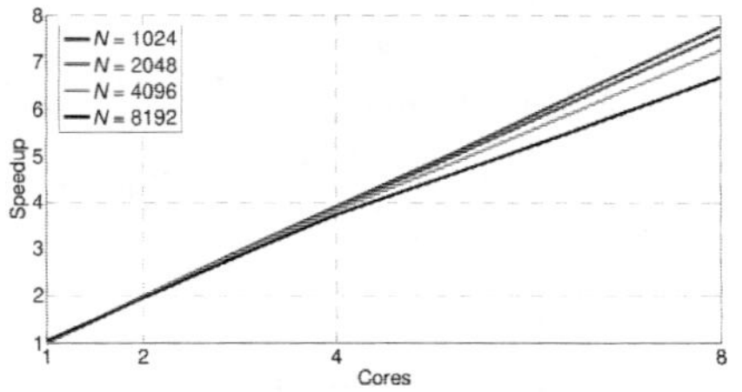

(d) Multicore, GPU DM, GSC.

Figure 3. APSO speedups.

5. Discussion

Although the speedups are consistent between trials using the GPU-accelerated DM and those using multicore parallelization (figures 3a, c), the timings for the GPU-accelerated method are much better than the multicore DM method (Table 1) for the high-dimensional composition, but are slightly worse for the GSC. From the profiling results, it is seen that computing $f(\mathbf{x})$ for GSC is the most intensive operation, and overwhelms the expensive DM calculation. Because GSC is 20D, the overhead incurred by data transfer slightly outweighs efficiency gains. For the 200D composition, the greater amount of work results in better speedup. In any case, parallelization in one form or another is required for DM. Additionally, although DM is a large part of APSO, there are other expensive components as well (e.g. complex cost function evaluation). Computation times for APSO were reduced from the single-core, no GPU configurations to the 8-core parallelized, with GPU-based DM calculation configurations by rates of 6.72 – 7.75 for GSC, and 6.49 – 6.85 for the composition function with cuBLAS matrix multiplication (see Table 1). Even better speedups are obtained when comparing the fully parallelized composition (with cuBLAS acceleration and GPU-based DM calculation) to the non-parallelized (single-core, no GPU) configuration, where speedups of 8.45 – 8.67 are achieved.

Cache size also has an effect on performance. Some speedup drops are seen in individual APSO components for moderate and high dimensions (see figures 2a-c), likely because the small cache on the

CPU cannot keep larger problems in cache, resulting in accesses to slower RAM. It is also seen that, in some cases, smaller N show a greater speedup than for larger populations (figures 2a-b). A plausible explanation is that as the size of the problem increases, less cache memory can be used, forcing access to higher, slower levels of cache, or even into RAM. However, the lower gains made by using multicore are justified because if the GPU is exclusively utilized, the CPU would be mostly idle. Furthermore, although the GPU has many cores, they are inherently weak and memory-limited vis-à-vis the CPU, which limits them to certain types of parallel problems. As access to RAM is very slow, required data may be preloaded, but only a small amount can fit into GPU memory.

Table 1. Mean (± std. dev.) optimization time (seconds) for cost functions by number of particles.

	Particles			
Cost Function / Configuration	1024	2048	4096	8192
GSC (1 CPU)	1050.37 ± 2.75	2125.61 ± 2.81	4329.69 ± 9.11	9004.92 ± 14.52
GSC (8 CPUs)	133.66 ± 0.28	273.73 ± 0.28	574.48 ± 0.81	1249.26 ± 1.17
GSC (8 CPUs-GPU DM)	135.57 ± 0.44	280.58 ± 0.57	597.20 ± 0.99	1340.92 ± 2.10
Speedup (8 CPUs/8 CPUs-GPU)	**7.86 / 7.75**	**7.76 / 7.58**	**7.54 / 7.25**	**7.21 / 6.72**
Comp. cuBLAS (1 CPU)	1149.52 ± 10.70	2454.29 ± 24.88	5483.02 ± 11.53	13460.55 ± 43.68
Comp. cuBLAS (8 CPUs)	178.87 ± 2.22	377.90 ± 3.80	844.23 ± 2.24	2067.58 ± 9.60
Comp cuBLAS (8 CPUs-GPU DM)	177.11 ± 0.50	374.80 ± 1.74	831.13 ± 5.97	1964.22 ± 8.75
Speedup (8 CPUs/8 CPUs-GPU)	**6.42 / 6.49**	**6.49 / 6.55**	**6.49 / 6.60**	**6.51 / 6.85**
Composition (1 CPU)	1535.99 ± 16.40	3206.76 ± 5.55	7060.97 ± 85.63	16588.32 ± 140.65
Composition (8 CPUs)	224.88 ± 1.02	471.10 ± 3.53	1028.84 ± 4.67	2434.49 ± 5.80
Composition (8 CPUs-GPU DM)	177.11 ± 0.50	374.80 ± 1.74	831.13 ± 5.97	1964.22 ± 8.75
Speedup (8 CPUs/8 CPUs-GPU)	**6.83 / 8.67**	**6.81 / 8.56**	**6.86 / 8.50**	**6.81 / 8.45**

6. Conclusion

The results presented in this paper are encouraging for applying heterogeneous computing to difficult global optimization problems using PSO. However, especially as evidenced in the GSC results, one cannot rely on GPU acceleration alone, and the key to increased multicore improvement is tighter memory-CPU integration [10]. Increasing CPU-GPU integration may reduce or eliminate the data transfer bottleneck. It is expected that in the near future, many-core chips and hybrid CPU/GPU processors, as well as greater exploitation of DSP and FPGA capabilities, will lead to further efficiency improvements in computationally-intensive applications, such as high-dimensional global optimization. These heterogeneous frameworks will also be instrumental in new "big-data" applications, so that "big-data supercomputing on a budget" will be more attainable.

Future work will include further improving the robustness and efficiency of PSO. For the lower-dimensional GSC problem, the effect of decreasing the population size and increasing the number of iterations to the overall efficiency will be assessed. For all problems, after 500 iterations, the population size can be decreased, or an efficient local method (e.g. Powell's method, multidirectional search) can be used to refine the best optimum found by APSO [26]. In addition, the intensive nested loop GSC cost function will be revisited for possible GPU execution. In future GPUs and other accelerators, it is expected that tighter accelerator/CPU connections will alleviate some of the bandwidth difficulties currently experienced in large problems. Finally, although PSO is robust for many classes of problems, many advanced techniques, including surrogate response surface methods and those using gradient information, will also be analyzed for efficiency gains using heterogeneous hardware, especially many-core systems with GPU or DSP acceleration.

Acknowledgments
The authors thank Dr. Renata Wachowiak-Smolíková and Devin Rotondo for helpful comments and criticisms. This work was supported by SHARCNET facilities (www.sharcnet.ca). The authors are supported by the Natural Sciences and Engineering Research Council of Canada (MPW, #386586-2011).

References
[1] Blum C and Merkle D 2008 *Swarm Intelligence: Introduction and Applications* (New York: Springer-Verlag)
[2] Engelbrecht A P 2006 *Fundamentals of Computational Swarm Intelligence* (New Jersey: John Wiley & Sons)
[3] Schutte J F and Groenwold A A 2005 *J Global. Optim.* **31** 93–108
[4] Schutte J F, Reinbolt J A, Fregly B J, Haftka R T and George A D 2004 *Int. J. Numer. Meth. Eng.* **61(13)** 2296–315
[5] Hung Y and Wang W 2012 *Optimization Methods and Software* **27** 33-51
[6] Cagnoni S, Bacchini A and Mussi L 2012 *Applications of Evolutionary Computation* **7248** 406-415
[7] Mussi L, Daolio F and Cagnoni S 2010 *Information Sciences* **181** 4642-57
[8] Wachowiak M P and Lambe Foster A E 2012 *J. Phys.: Conf. Ser.* **385**
[9] Slabaugh G, Boyes R and Yang X 2010 *IEEE Signal Processing Magazine* **27-2** 134-8
[10] Moore S K 2008 *IEEE Spectrum* **Nov** 15
[11] Moore S K 2011 *IEEE Spectrum* (New York: Institute of Electrical and Electronics Engineers) **Jan** 40
[12] Kindratenko V V and Trancoso P 2011 *Computing in Science & Engineering* **13** 92-5
[13] Schnabel R 1995 *Parallel Computing* **21** 875-905
[14] Shi X, Liang Y, Lee H, Lu C and Wang L 2005 *Information Processing Letters* **93** 255-61
[15] Lovbjerg M, Rasmussen T K and Krink T 2001 *Proc. 3rd Genetic Evolutionary Computation Conf. (California)* 469-76
[16] Veeramachaneni K, Peram T, Mohan C and Osadiciw L A 2003 *Proc. Genetic and Evolutionary Computation Conf. 2003 (Illinois)* 110-21
[17] Montes De Oca M A, Stutzle T, Birattari M and Dorigo M 2009 *IEEE Transactions on Evolutionary Computation* **13** 1120-32
[18] Banks A, Vincent J and Anyakoha C 2008 *Natural Computing* **7.1** 109–24
[19] Nickabadi A, Ebadzadeh M M and Safabakhsh R 2011 *Applied Soft Computing* **11** 3658-70
[20] Zhan Z-H, Zhang J, Li Y and Chung H S-H 2009 *IEEE Transactions on Systems, Man, and Cybernetics, Part B (Cybernetics)* **39.6** 1362-81
[21] Liang J, Ponnuthurai J, Suganthan N and Deb K 2005 *Proc. 2005 IEEE In Swarm Intelligence Symposium (California)* 68-75
[22] Mathias K E, Whitley D D, Stork C and Kusuma T 1994 *Proc. of the First IEEE Conf. on IEEE World Congress on Computational Intelligence (Florida)* **1** 356-61
[23] Whitley D, Lunacek M and Sokolov A 2006 *Lecture Notes in Computer Science* **4193** 988-997
[24] Moore R 2011 *IS&T/SPIE Electronic Imaging* (pp. 78720L-78720L)
[25] Chang D, Jones N A, Li D, Ouyang M, Ragade R K 2008 *Proc. of the IASTED Int. Symposium on Computational Biology and Bioinformatics* 278-283.
[26] Wachowiak M P and Peters T M 2006 *IEEE Trans Inform. Tech Bio* **10(2)** 344-53

High Performance Computing Symposium 2013 (HPCS 2013) IOP Publishing
Journal of Physics: Conference Series **540** (2014) 012008 doi:10.1088/1742-6596/540/1/012008

Modeling a distributed environment for a petroleum reservoir engineering application with software product line

Rafael de Faria Scheidt, Patrícia Vilain, M. A. R. Dantas
Research Laboratory of Distributed Systems (LaPeSD)
Department of Informatics and Statistic (INE)
Federal University of Federal Santa Catarina (UFSC)
Florianópolis, SC - Brazil

E-mail: {rfscheidt, vilain, dantas}@inf.ufsc.br

Abstract. Petroleum reservoir engineering is a complex and interesting field that requires large amount of computational facilities to achieve successful results. Usually, software environments for this field are developed without taking care out of possible interactions and extensibilities required by reservoir engineers. In this paper, we present a research work which it is characterized by the design and implementation based on a software product line model for a real distributed reservoir engineering environment. Experimental results indicate successfully the utilization of this approach for the design of distributed software architecture. In addition, all components from the proposal provided greater visibility of the organization and processes for the reservoir engineers.

1. Introduction

The reuse of elements and components inside projects with a similar domain scope in software engineering and object oriented programming approaches is a common practice. This effort is related to the reduce of costs, time and productivity [1]. Aiming to reach these goals several techniques were proposed, such as: domain engineering, frameworks, patterns, software architecture and also development based on components [2].

The software product line (SPL) is a characteristic approach which was conceived to organize systematically and also provide a predictable fashion to software reuse based on the same product domain. As mentioned in [3], the SPL technique appears from products or similar researches in the same domain to a common business area. In accordance to the Software Engineering Institute (SEI) [1], the SPL paradigm represents software systems which share a set of common and controlled characteristics. These satisfies a special market segment and are developed from key artifacts (core assets), in a pre-defined form.

Several organizations, when implementing their software domain applications, adopt some particular characteristics to differentiate one specific product [18]. Thus, these organizations are always seeking for processes and approaches that could help in the reuse of the main characteristics from a specific product.

This paper presents a research work which is depicted by a design and an implementation of software product, for a reservoir engineering environment, adopting the software product line (SPL) paradigm.

The contribution is characterized for considering a real application, which was being developed at the Research Laboratory of Distributed Systems (LaPeSD) at Federal University of Santa Catarina. In this group, with large experience in developing distributed applications, similar to other peers groups, when a new version of a product was demanded several artifacts were reused, albeit observing an evolution of some characteristics. Therefore, it is possible to verify the existence of core architecture with some components variants.

In addition, it was frequently the case where the client asks for a specific feature and afterwards asks to remove this facility. After sometime, the client claimed again for that specific module, but now the system could not support it. In other words, in both sides, clients and developers, there were no culture of reuse. Therefore, the utilization of the SPL approach may improve operations such as inclusion and remove of elements from the distributed architecture. As a result, it was expected to achieve a better level of flexibility, maintenance and product evolution.

The paper is organized as follows. Related works are presented in section 2. In section 3, we show the conventional distributed configuration utilized for the development that is contrasted with the proposal considering the use of the SPL technique. Empirical results are presented in section 4. Finally, in section 5 we presented conclusions and future work related to this research.

2. Related Work

In this section, it is presented two research projects in which distributed configurations were conceived as an intrinsic feature together with the SPL approach. These two related works were chosen because it was possible to have more access to information from these proposals.

2.1. UBÁ

The UBÁ proposal [14] aims to build SPL architecture to help the development of middleware systems to the grid computing configurations. The approach is used to minimize the complexity related to this configuration. Examples of some facilities created are: the coordination for large scale resource sharing; Quality of Service (QoS) provision; support of a large number of heterogeneous devices; geographic dispersion of resources, people and applications.

This research was characterized through the SPL instantiation of the proposal, thus creating a middleware for grid computing which provides their main facilities. In addition, it was built a structure that utilizes the API from the created middleware. The configuration was tested in different scenarios, executing successfully one application in these different environments.

2.2. Systemized Reuse of SPL in Financial Segment

A study presented in [16] covers a system reuse, through the utilization of the SPL, in the financial segment. This document illustrates studies referring to the five major Brazilian banks. The key element of this research is to present how the reuse is adopted inside financial sector and how this practice contributes to the success of software projects.

An interesting example that pointed out the necessary use of reuse approach is illustrated by a specific bank that executes one modification in ordinaries accounts and it was required modification into 72 systems. This problem was identified as a lack of planning in the reuse inside those systems.

2.3. Observed Aspects

The UBÁ approach focuses its development for a specific application domain as a grid computing middleware. In other words, domain engineering for a broadly distributed system is not covered.

On the other hand, the research project presented in [16] illustrates that the SPL approach is not a key element for those organizations. The functions of coupling and maintenance for the applications do not consider a software product line, and also it exist a lack of reuse as macro paradigm.

Understanding the former two contributions, which could have some intersection aspects with the target configuration of our research, it was essential to conceive a differentiated contribution considering a more generic distributed system environment.
Therefore, an effort in which can be possible represent the domain engineering and application instantiation for distributed system configurations, may answer our research question. As result, our proposal should conceive to design and implement a software environment approach based on the SPL paradigm.

It was not found in the literature a similar proposal, considering a real case study, in which a generic distributed system had a support from the SPL with a high level of artifacts reuse.

3. Proposal

In this section, it is described the research proposal in terms of design and implementation of the software product line approach for a real distributed system configuration. Thus, the section shows how was the old development process for configuration, the validation for the job management system and how was the reformulation from the old-fashioned to the new paradigm.

3.1. The Original Development Environment

The development of an application for any large distributed system configuration can be seen as a complex process. As mentioned in [17], heterogeneity in terms of hardware, operating systems and resource management systems (RMS) are three examples of key aspects to be considered for any distributed project. As a result, elements such as graphic interfaces, network interfaces and also programming middleware software packages are important to be analyzed.

The main contribution of this research work is to apply the SPL Smarty process in a real distributed system project, which is being developed at the Research Distributed System Laboratory (LaPeSD), at the Federal University of Santa Catarina (UFSC), in collaboration with a petroleum engineering company.

As a result, the proposal is to contribute for the systematization of future developments considering artifacts, components and architecture reuse.

In this context, each element (e.g. GUI, interfaces and RMS) could be translated to components forming a common architecture known as domain engineering (DE). Afterwards, for any new product will be possible to have an instance analysis for this new application engineering (AE) architecture, designing a specialization for those no common parts. Through the utilization of the SPL approach, it is possible to have a systematic process, adopting the reuse paradigm and also a planning procedure to add these new components to the architecture. Important to observe, these features will be added into the configuration without an extra effort.

However, to reach level of abstractions mentioned before, it was required changes in the actual architecture. The original development environment was characterized as a tightly coupled configuration and it was difficult apply the reuse and componentization techniques. Figure 1 shows a macro idea, on how was the configuration.

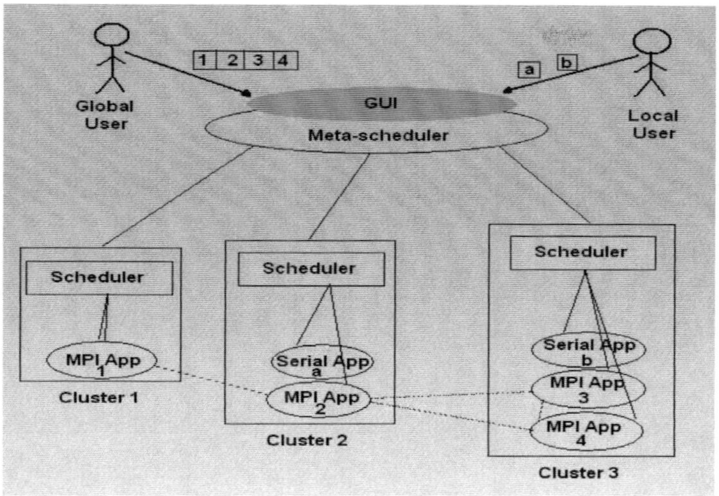

Figure 1. Macro view of the original development environment.

A meta-scheduler element has a key importance in the distributed configuration. This component provides the interface between requirements from a user to a scheduler (RMS), in a specific cluster configuration. Usually, the meta-scheduler is coined as the middleware of the configuration. The main reason for this is based on the fact that each cluster has its own scheduler (e.g. Condor, OpenPBS and LSF). Thus, the meta-scheduler is obliged to know each API from all schedulers and afterwards realize all communications.

In a case of a new cluster added to the configuration, the middleware is informed and all elements necessary to this communication should be provided by the middleware. Location, ports, components and general descriptions are examples of elements necessary for all communications. Therefore, this is one new component that should be aggregated to the middleware environment, and which also should be available to interoperate with other existing components.

This contribution can be considered differentiated, because it proposes an approach which can be translated in a systematic method, reusable and with previous planning using the SPL paradigm. As a result, it will be possible to add components to the original distributed system architecture without large changes for the middleware environment.

To reach the necessary transparency, it is necessary to process some changes into the original middleware architecture. In other words, our contribution presents a new architecture approach with a reusable function and components with variations points that were previously defined. These definitions were designed foreseeing new functionalities, components and schedulers.

3.2. The New Development Environment

The ordinary process of software product line is shown in figure 2. In each phase of the process, artifacts are produced and stored inside a SPL repository.

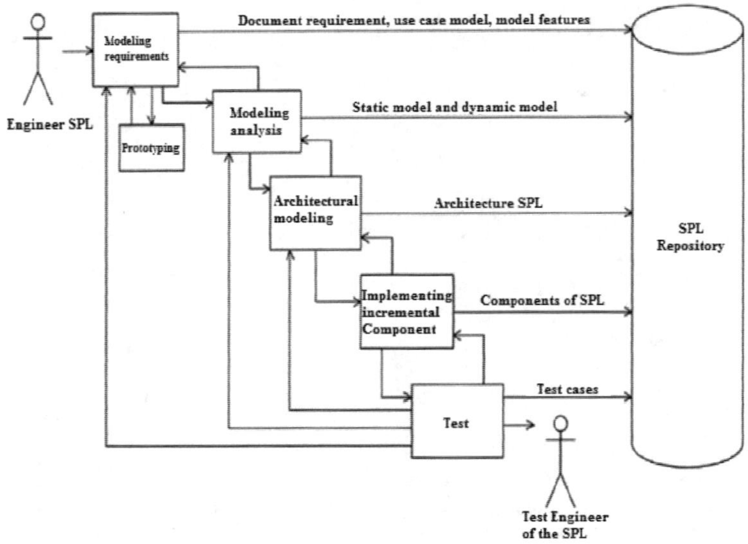

Figure 2. SPL developing process.

The process starts with the requisite modeling technique which comprises the model elaboration and requisite document, use cases models and a feature model. Use case models define SPL functions requisites in terms of actors and use cases. It is also possible to realize the similarity and variability model, in this research study, from the Smarty approach. Inside the features model, each item is a requisite that defines one similarity or variability which could be translated into parameters.

In the requisite analysis phase, all SPL functional and non-functional requisites are specified. These requisites are verified in the next steps to check their relations. Thus, it is established restrictions through pre-requisites. Found all pre-requisites, these will guide the direction of the development, for build the architecture and its validation.

3.2.1. Functional Requisites

1. RF001 – Advanced Resource Reservation.

Description: The system should be able to provide an advanced resource reservation function for a task to be executed in the future with the required resources.

2. RF002 – Immediate Resource Reservation.

Description: The system should be able to provide immediate resource reservation function

3. RF003 – Resource Reservation Based on Characteristics.

3.2.2. Non-Functional Requisites

4. RNF001 – Add/Remove enhancements or functionalities with impact the actual structure.

Description: The structure should be able to support the add and remove functions to improve existing functionalities of the architectures. It is also possible to add functionalities and components to the environment without any prejudice for users and services. Example of such prejudice is the communication between services.

5. RNF002 – Heterogeneous Services.

The architecture should support any operating system and hardware

Departing from functional and non-functional requisites it is possible to identify use cases that exist in the domain model. Before the execution of the use cases, it is imperative to identify all possible actors in the system as illustrated in figure 3.

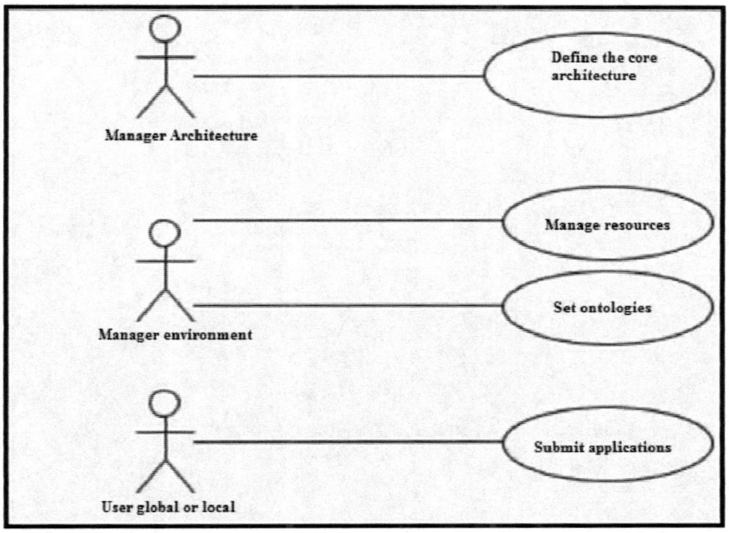

Figure 3. Actors from the use case model.

All use cases require one refined action and specific actions from each actor, which should be explained in their user cases.

The use case evolved in figure 4 presents innumerous variance points such as: "Define the main architecture"; "Define GUI"; "Define meta-scheduler"; "Define LaPeSD products"; "Define RMS". A special fact to be considered is the occurrence of obliged activities as "Define rules" and optional activities such as "Allows dynamic changes".

In addition to the variation point "Define meta-scheduler", there are two variants with stereotypes "alternative XOR". This indicates that only one option is possible to be chosen. The specification exists as exclusive alternative relation inside the SMarty approach.

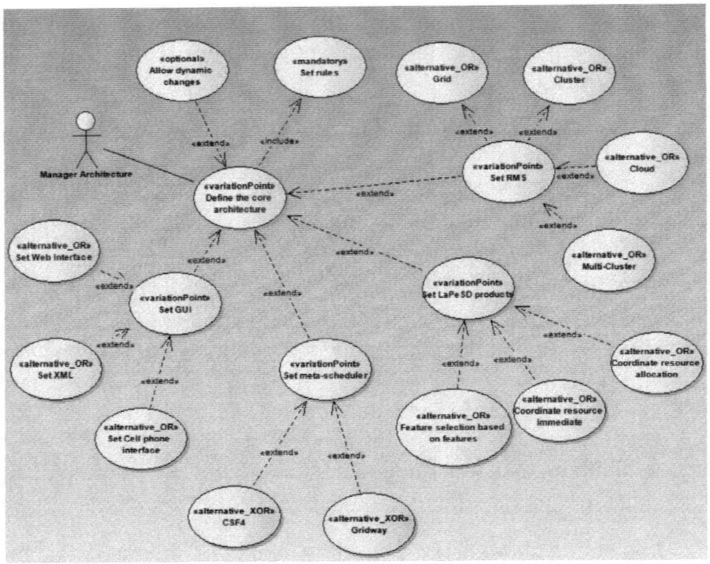

Figure 4. Variability.

The architectural modeling is an important phase in the SPL approach. This phase is characterized by a specification and definition of the architecture based on components. In this case, the analysis model is addressed to the architecture model taking into consideration software patterns.

Figure 5 shows the proposed architecture after applying the SPL approach for the distributed system environment. The vision layer is composed by the communication and job submission interfaces. In addition, a user may select resources and necessary attributes. After this layer, it is presented the main variation point from the architecture. This communication interface adopts as variation points some general Communicators, such as Condor, SGE and LSF. These items if instantiated in a domain engineering, it will be implemented as their RMS specifications. This new feature allows that new RMS may be added to the architecture. The implementation of these interfaces is a product from the LaPeSD.

4. Experimental Results

In this section it is presented some experimental results, after adopting the SPL approach in the project. The evaluation of proposal was realized through study case in the real project from the laboratory. Therefore, all information related to the architecture, components and artifacts are real.

The complexity of the environment was bypassed, first conceiving a model for the graphical user interface in which was possible to submit jobs. As a result, our effort was to add a special focus in the development, implementation and validation of this layer, due to its key importance.

The access to the environment was conceived through the use case from the interface layer. It was utilized variation points and delimitated variances using the SPL. The SmartyProcess approach was the choice which is represented by: web interface, cellular interface and WebService. These elements will be the family products from the LaPeSD, considering the layer interface with the user.

In addition to previous interfaces, it was designed a communication interface abstraction with several RMS that exist and could be possible for the architecture. Condor, SGE and LSF communicator and variance were implemented. Each communicator specifies the communication form inside each RMS, where each one has its commands.

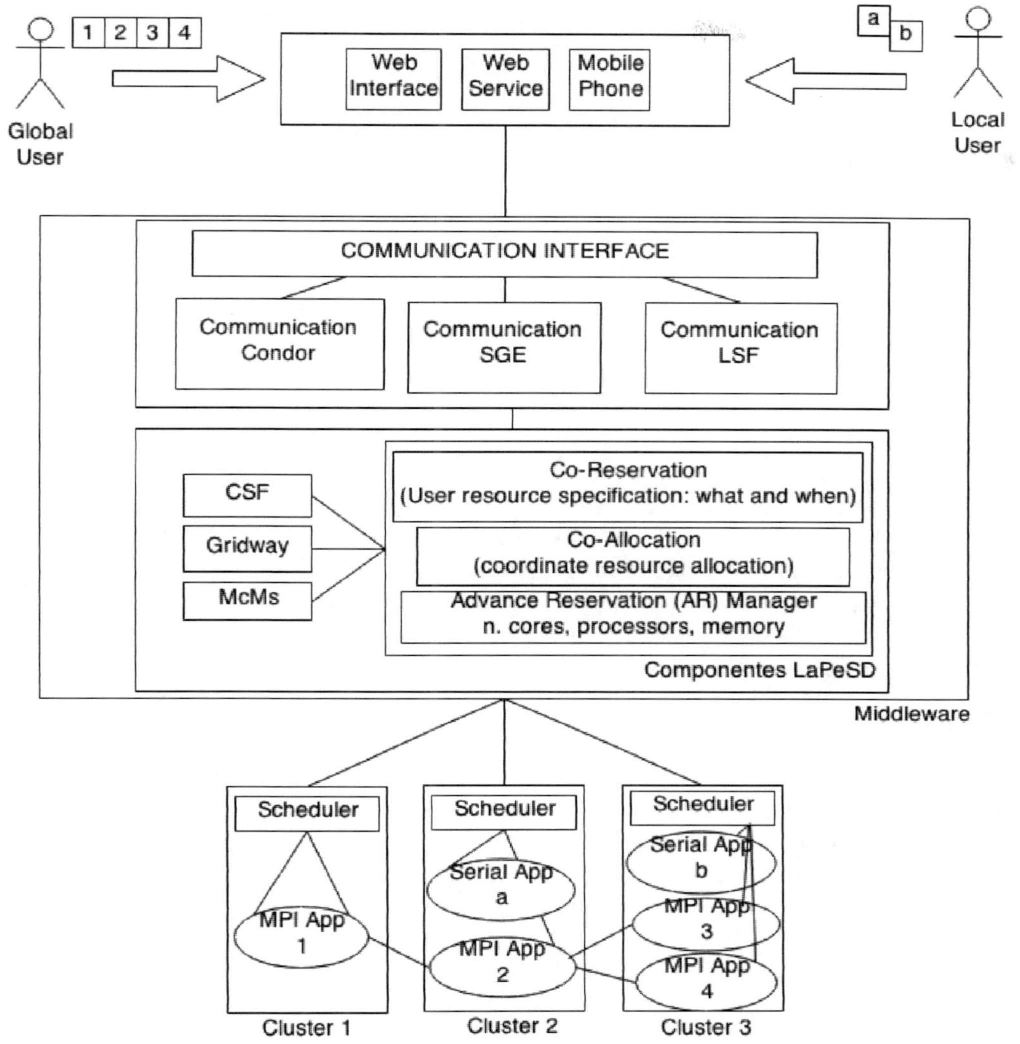

Figure 5. Proposed architecture.

The communication procedure was decoupled from the interface and created one structured in edge of the middleware. As a result, components from the interface were turned into components and were open to new interface designs or evolutions. An example, it could be seen in the future a new graphical interface considering dynamic resource management adopting a fuzzy logic paradigm to hardware and software selection.

A pool was realized among experts from the laboratory and it was possible to map messages that are usually displayed to end users. These were implemented in the communication interface for the graphical interfaces for each communicator as shown in table I.

It this phase of our work it was created a webservice, which made possible any implementation in the upper layer (or GUI). Therefore, it was possible to utilize all messages from table I, where a Global or a Local User was able to submit jobs, process some management tasks or a RMS.

Table 1. EXISTING MESSAGES IN THE RMS COMMUNICATION.

Message	Function	SGE Command	Condor Command
submitJobSynch()	This method is called for a syn submission	Qsub	Condor_submit
delJob()	Use to cancel/delete jobs	Qdel	Condor_rm
getJobs()	List job in execution mode	Qstat	Condor_q
listNodes()	List information status about nodes and execution nodes	Qhost	Condor_status

Afterwards, the communication interface which is represented by the Communication class inside the webservice, receive a request this is forwarded to the CommunicationServiceImpl class. This latter will verify which RMS will be manipulated. The submission through the user GUI requires an indication of a RMS. As soon as a requests arrives at the webservice, following the request it knows which class will be invoked, i.e. CondorService or SGEService. In case that the class CondorService was chosen, this component will be communicating to the Local RMS and will be processed their own specifications and requirements. As an example, if a job submission is required the specific command of the Condor RMS will be executed, without the necessary knowledge of any syntax from the end user.

5. Conclusions and Future Work

In this research work it was presented a design and implementation of the software project line approach in a real distributed system project. The goal was to provide in a systematic fashion components, also to create domain engineering. The aim was successfully reached.

Before the present effort, the original development environment was characterized as a tightly coupled configuration and it was difficult apply the reuse and componentization techniques. One frequently occurrence was that client asks for a specific feature and afterwards asks to remove this facility. After sometime, the client claimed again for that specific module, but now the system could not support it. In other words, in both sides, clients and developers, there were no culture of reuse. Therefore, the utilization of the SPL approach has improved operations such as inclusion and remove of elements from the distributed architecture. This objective was reached through a layer, where all requests were pre-processed and redirected to an appropriate RMS module. As a final result, the conceived SLP environment provides an enhancement artifact reuse approach where errors were lower and software conformity was higher. In addition, characteristics such as an easier maintenance and less time to develop features were verified in the new configuration.

As a future research plan, we are going to implement the SPL approach in ubiquitous part of the software environment. In addition, we are going to develop a set of tools to provide some support and automatization for the design of applications. It will be also important conceive and implement a tool which may provide an enhanced option to select components and variants of distributed systems, thus improving applications creation.

6. References

[1] SEI, Software Engineer Institute [online]. Available: http://www.sei.cmu.edu/, 2013
[2] Gimenes, I. M. S.; Travassos, Guilherme Horta. The Product Line Approach to Software Development. Ingrid Jansch Porto. Porto Alegre, Brazil, 2002.

Journal of Physics: Conference Series **540** (2014) 012008 doi:10.1088/1742-6596/540/1/012008

[3] Vasconcelos, A. One approach to support the creation of reference architectures based on domain analysis of Legacy Systems. Phd Thesis, Federal University Federal of Rio de Janeiro, Rio de Janeiro, Brazil, 2007.
[4] Dantas, M. High Performance Distributed Computing: networks, clusters and computational grids.Rio de Janeiro: Axcel Books, Brazil, 2005.
[5] De Rose, C. A. F.; Ferreto, T. C., Improving Performance Analysis Using Resource Management Information, Springer Berlin/Heidelberg, 2003.
[6] Buyya, R. Grid Computing Info Centre (GRID Infoware). Available: http://gridcomputing.com, 2011.
[7] Colvero, T. A.; Dantas, M.; Cunha, D. P. da. Environments clusters and computational grids: features, facilities and challenges. Criciúma: Unesc, Santa catarina, Brazil, 2005.
[8] Yeo, C. S. et al. Cluster Computing: High-Performance, High-Availability, and High-Throughput Processing on a Network of Computers. 2006
[9] Clements, P.; Northrop, L. Software Product Lines: Practices and Patterns. Boston, MA, USA: Addison-Wesley Longman Publishing Co., Inc., 2001.
[10] Northrop, L. M. SEIs Software Product Line Tents. IEEE Software, v. 19, n. 14, p. 32-41, 2002.
[11] Pohl, K.; Bockle, G.; Linden, F. J. V. D. Software Product Line Engineering: Fundations, Principles and Techniques. Secaucus, NJ, USA: Springer-Verlag, 2005.
[12] Oliveira Junior, E. A.; Gimenes, Itana M. S.; Maldonado, José C. Systematic Management of Variability in UML-based Software Product Lines. Journal of Universal Computer Science, p. 1-20, 2010.
[13] Oliveira, A. Edson SystEM-PLA: A Method for Evaluating Product Line Architecture Software Based on UML, Instituto de Ciências Matemáticas e Computação, State University of São Paulo, Brazil, Ph.D. Thesis, 2011.
[14] Oliveira, D.; Rosa, N. Ubá: A Software Product Line Architecture for Grid-Oriented Middleware. Informatic Center, Fed. Univ. of Para, Santarem, Brazil.33rd Annual IEEE International Computer Software and Applications Conference, 2009.
[15] Bosch, J. Preface. In: The 2nd Groningen Workshop on Software Variability Management: software product families and populations, 2004, Groningen, The Netherlands. Proceedings. Groningen, The Netherlands, 2004.
[16] Reinehr, Sheila dos Santos. Systematic reuse of software and Software Product Lines in the financial sector: a case study in Brazil.2008. 327 f. Escola Politécnica, Programa de Pós-Graduação em Engenharia de Produção, Phd Thesis, São Paulo, Brazil,2008.
[17] Silva A P C and Dantas M A R 19th International Symposium on Computer Architecture and High Performance Computing, (SBAC-PAD'07) 143-150, 2007.
[18] Sugumaran, V.; Park, S.; Kang, K. C. Software Product Line Engineering. Communications of the ACM, v. 49, n. 12, 2006.

Institute of Physics
Dirac House, Temple Back
Bristol BS1 6BE UK

CURRAN ASSOCIATES INC
proceedings.com